Student Solutions Manual to accompany Nakos and Joyner's

AF709879

Linear Algebra with Applications

George Nakos
U.S. Naval Academy

Brooks/Cole Publishing Company
I(T)P® An International Thomson Publishing Company
Pacific Grove • Albany • Belmont • Bonn • Boston • Cincinnati • Detroit • Johannesburg • London
Madrid • Melbourne • Mexico City • New York • Paris • Singapore • Tokyo • Toronto • Washington

A GARY W. OSTEDT BOOK

Publisher: *Gary W. Ostedt*
Editorial Associate: *Carol Ann Benedict*
Marketing Manager: *Heather Woods*
Production Editor: *Mary Vezilich*
Printing and Binding: *Webcom*

For more information, contact:

BROOKS/COLE PUBLISHING COMPANY
511 Forest Lodge Road
Pacific Grove, CA 93950
USA

International Thomson Publishing Europe
Berkshire House 168-173
High Holborn
London WC1V 7AA
England

Thomas Nelson Australia
102 Dodds Street
South Melbourne, 3205
Victoria, Australia

Nelson Canada
1120 Birchmount Road
Scarborough, Ontario
Canada M1K 5G4

International Thomson Editores
Seneca 53
Col. Polanco
11560 México, D. F., México

International Thomson Publishing Japan
Hirakawacho Kyowa Building, 3F
2-2-1 Hirakawacho
Chiyoda-ku, Tokyo 102
Japan

International Thomson Publishing Asia
221 Henderson Road
#05-10 Henderson Building
Singapore 0315

International Thomson Publishing GmbH
Königswinterer Strasse 418
53227 Bonn
Germany

Printed in Canada

5 4 3 2

ISBN 0-534-95529-0

Στους γονεις μου

Preface

This *Student Solutions Manual* contains solutions to all odd-numbered exercises in the text *Linear Algebra and Applications*, by George Nakos and David Joyner. Solutions to the projects are not included. Also, solutions to computer exercises are not included, as most of these appear in the main text.

This manual supplements the main text. Its solutions are usually detailed enough to help the student practice with calculations, concepts, applications, and theory.

Any feedback and corrections of errors are greatly appreciated. Please either contact me at the address below, or write to the publisher: Brooks/Cole Publishing Company, 511 Forest Lodge Road, Pacific Grove, CA 93950.

I would like to thank Gary Ostedt and Carol Benedict for the opportunity to publish with Brooks/Cole, and for all the help and support for the past few years. I also want to thank some of the most supportive friends and colleagues: Dr. Robert M. Williams and the U.S. Naval Academy professors Mike W. Chamberlain, Carol G. Crawford, James M. D'Archangelo, Reza Malek-Madani, Geoffrey L. Price, and Peter R. Turner.

I dedicate this book to my family: my parents Constantine and Helen, my wife Debra, and my sons Constantine and David.

George Nakos
Mathematics Department, M/S 9E
U.S. Naval Academy
572 Holloway Road
Chauvenet Hall
Annapolis, MD 21402

Contents

Exercises 1.1

The notation $A \sim B$ is used to separate two consecutive stages A and B in the elimination process.

1. Equations (a)–(e) are linear. Equation (f) is nonlinear. Equation (c) is the only homogeneous equation.

3. Equations (a)–(e) are linear.

Equation	Variables	Leading variable	Free variables
(a)	x, y	y	x
(b)	x, y	None	x, y
(c)	x, y, z	x	y, z
(d)	x, y, z	x	y, z
(e)	x, y, z, w, t	x	y, z, w, t

5. Only S is on the plane.

7. (a) The equation can be rewritten in the form $(a^2 - 4)x = a + 2$. If $a = -2$, then $0x = 0$, so there are infinitely many solutions. If $a = 2$, then $0x = 4$, so there are no solutions. If $a \neq \pm 2$, then there is the unique solution $x = 1/(a - 2)$.

(b) If $a = \pm 2$, we get $0x = 3$, so there are no solutions. If $a \neq \pm 2$, we have the unique solution $x = 3/(a^2 - 4)$.

(c) If $a = \pm 2$, we get $0x = 0$, so there are infinitely many solutions. If $a \neq \pm 2$, there is the unique solution $x = 0$.

(d) If $a = 0$, we get $0x = 0$, so there are infinitely many solutions given by $x = r$, $y = t$. If $a \neq 0$, there are infinitely many solutions given by $x = 3 + ar$, $y = r$.

9. The standard form is

$$\begin{array}{rrrrrcr} 2x & & +4z & & & = & -1 \\ -x & & +2z & & +2w & = & 2 \\ -2x & & -z & & +3w & = & -3 \\ & y & +z & +t & -w & = & 4 \end{array}$$

(a)–(c) The coefficient matrix and the vector of constants are

$$\begin{bmatrix} 2 & 0 & 4 & 0 & 0 \\ -1 & 0 & 2 & 0 & 2 \\ -2 & 0 & -1 & 0 & 3 \\ 0 & 1 & 1 & 1 & -1 \end{bmatrix}, \quad \begin{bmatrix} -1 \\ 2 \\ -3 \\ 4 \end{bmatrix}$$

The augmented matrix is

$$\begin{bmatrix} 2 & 0 & 4 & 0 & 0 & : & -1 \\ -1 & 0 & 2 & 0 & 2 & : & 2 \\ -2 & 0 & -1 & 0 & 3 & : & -3 \\ 0 & 1 & 1 & 1 & -1 & : & 4 \end{bmatrix}$$

(d) The associated homogeneous system is

$$\begin{array}{rrrrrcl} 2x & & +4z & & & = & 0 \\ -x & & +2z & & +2w & = & 0 \\ -2x & & -z & & +3w & = & 0 \\ & y & +z & +t & -w & = & 0 \end{array}$$

11. If $x_5 = s$, then the last equation implies that $x_4 = s/2$. If $x_6 = r$, then the second equation implies that $x_3 = 2r$. If $x_2 = t$, then the first equation implies that $x_1 = -2t - 2r - s$. So, the general solution is $(-2t - 2r - s, t, 2r, s/2, s, r)$ for any $r, s, t \in \mathbf{R}$.

13. By the second equation, if $x_6 = r$, then $x_5 = r$. Therefore, $x_3 = r$, by the third equation. If $x_4 = s$ and $x_2 = t$, then $x_1 = 1 + t - r + 5s - 6r + r$, by the first equation. Hence, the general solution is $x_1 = 1 + t - 6r + 5s$, $x_2 = t$, $x_3 = r$, $x_4 = s$, $x_5 = r$ and $x_6 = r$ for all r, s, t.

15. Solving the system yields $x = 1/6$ and $y = 5/6$. So, the intersection is $P(1/6, 5/6)$.

17. In the second system: the first equation is a multiple of the first equation of the first system. And the last equation is the sum of all 3 equations of the first system.

19. We have

$$\begin{bmatrix} 0 & 1 & 2 & 6 \\ 3 & -3 & -3 & -15 \\ 1 & 3 & 3 & 11 \end{bmatrix} \sim \begin{bmatrix} 1 & 3 & 3 & 11 \\ 3 & -3 & -3 & -15 \\ 0 & 1 & 2 & 6 \end{bmatrix} \sim \begin{bmatrix} 1 & 3 & 3 & 11 \\ 0 & -12 & -12 & -48 \\ 0 & 1 & 2 & 6 \end{bmatrix} \sim$$

$$\begin{bmatrix} 1 & 3 & 3 & 11 \\ 0 & 1 & 1 & 4 \\ 0 & 1 & 2 & 6 \end{bmatrix} \sim \begin{bmatrix} 1 & 3 & 3 & 11 \\ 0 & 1 & 1 & 4 \\ 0 & 0 & 1 & 2 \end{bmatrix} \sim$$

$$\begin{bmatrix} 1 & 3 & 0 & 5 \\ 0 & 1 & 0 & 2 \\ 0 & 0 & 1 & 2 \end{bmatrix} \sim \begin{bmatrix} 1 & 0 & 0 & -1 \\ 0 & 1 & 0 & 2 \\ 0 & 0 & 1 & 2 \end{bmatrix}$$

Hence, the system is consistent with solution $x = -1, y = 2, z = 2$.

21. We have

$$\begin{bmatrix} 3 & 1 & 3 & 15 \\ -1 & 3 & -1 & -5 \\ 2 & 4 & 2 & 9 \end{bmatrix} \sim \begin{bmatrix} 3 & 1 & 3 & 15 \\ 0 & \frac{10}{3} & 0 & 0 \\ 0 & \frac{10}{3} & 0 & -1 \end{bmatrix} \sim \begin{bmatrix} 3 & 1 & 3 & 15 \\ 0 & \frac{10}{3} & 0 & 0 \\ 0 & 0 & 0 & -1 \end{bmatrix}$$

The last equation is impossible, hence the system is inconsistent.

23. We have

$$\begin{bmatrix} 1 & 1 & -3 & 2 \\ -3 & 1 & 1 & 6 \end{bmatrix} \sim \begin{bmatrix} 1 & 1 & -3 & 2 \\ 0 & 4 & -8 & 12 \end{bmatrix} \sim$$

$$\begin{bmatrix} 1 & 1 & -3 & 2 \\ 0 & 1 & -2 & 3 \end{bmatrix} \sim \begin{bmatrix} 1 & 0 & -1 & -1 \\ 0 & 1 & -2 & 3 \end{bmatrix}$$

Hence, the system is consistent with general solution $(r - 1, 2r + 3, r)$, $r \in \mathbf{R}$.

25. We have

$$\begin{bmatrix} 0 & 3 & 1 & -1 & 3 \\ 1 & 1 & -2 & 0 & 6 \\ -2 & 1 & 2 & -1 & 9 \end{bmatrix} \sim \begin{bmatrix} 1 & 1 & -2 & 0 & 6 \\ 0 & 3 & 1 & -1 & 3 \\ -2 & 1 & 2 & -1 & 9 \end{bmatrix} \sim \begin{bmatrix} 1 & 1 & -2 & 0 & 6 \\ 0 & 3 & 1 & -1 & 3 \\ 0 & 3 & -2 & -1 & 21 \end{bmatrix} \sim$$

$$\begin{bmatrix} 1 & 1 & -2 & 0 & 6 \\ 0 & 3 & 1 & -1 & 3 \\ 0 & 3 & -2 & -1 & 21 \end{bmatrix} \sim \begin{bmatrix} 1 & 1 & -2 & 0 & 6 \\ 0 & 3 & 1 & -1 & 3 \\ 0 & 0 & -3 & 0 & 18 \end{bmatrix} \sim \begin{bmatrix} 1 & 1 & -2 & 0 & 6 \\ 0 & 3 & 1 & -1 & 3 \\ 0 & 0 & 1 & 0 & -6 \end{bmatrix} \sim$$

$$\begin{bmatrix} 1 & 1 & -2 & 0 & 6 \\ 0 & 3 & 0 & -1 & 9 \\ 0 & 0 & 1 & 0 & -6 \end{bmatrix} \sim \begin{bmatrix} 1 & 1 & 0 & 0 & -6 \\ 0 & 3 & 0 & -1 & 9 \\ 0 & 0 & 1 & 0 & -6 \end{bmatrix} \sim$$

$$\begin{bmatrix} 1 & 1 & 0 & 0 & -6 \\ 0 & 1 & 0 & -\frac{1}{3} & 3 \\ 0 & 0 & 1 & 0 & -6 \end{bmatrix} \sim \begin{bmatrix} 1 & 0 & 0 & \frac{1}{3} & -9 \\ 0 & 1 & 0 & -\frac{1}{3} & 3 \\ 0 & 0 & 1 & 0 & -6 \end{bmatrix}$$

So, the system is consistent with general solution $\left(-\frac{r}{3} - 9, \frac{r}{3} + 3, -6, r\right)$, $r \in \mathbf{R}$.

27. We have

$$\begin{bmatrix} 1 & 1 & 0 & 0 & 1 \\ 0 & 1 & 1 & 0 & 1 \\ 0 & 0 & 1 & 1 & 1 \\ 1 & 0 & 0 & 1 & 1 \end{bmatrix} \sim \begin{bmatrix} 1 & 1 & 0 & 0 & 1 \\ 0 & 1 & 1 & 0 & 1 \\ 0 & 0 & 1 & 1 & 1 \\ 0 & -1 & 0 & 1 & 0 \end{bmatrix} \sim \begin{bmatrix} 1 & 1 & 0 & 0 & 1 \\ 0 & 1 & 1 & 0 & 1 \\ 0 & 0 & 1 & 1 & 1 \\ 0 & 0 & 1 & 1 & 1 \end{bmatrix} \sim$$

$$\begin{bmatrix} 1 & 1 & 0 & 0 & 1 \\ 0 & 1 & 1 & 0 & 1 \\ 0 & 0 & 1 & 1 & 1 \\ 0 & 0 & 0 & 0 & 0 \end{bmatrix} \sim \begin{bmatrix} 1 & 1 & 0 & 0 & 1 \\ 0 & 1 & 0 & -1 & 0 \\ 0 & 0 & 1 & 1 & 1 \\ 0 & 0 & 0 & 0 & 0 \end{bmatrix} \sim \begin{bmatrix} 1 & 0 & 0 & 1 & 1 \\ 0 & 1 & 0 & -1 & 0 \\ 0 & 0 & 1 & 1 & 1 \\ 0 & 0 & 0 & 0 & 0 \end{bmatrix}$$

Hence, the system is consistent with general solution $(1 - r, r, 1 - r, r)$, $r \in \mathbf{R}$.

29. We have

$$\begin{bmatrix} 3 & 2 & 1 & 39 \\ 2 & 3 & 1 & 34 \\ 1 & 2 & 3 & 26 \end{bmatrix} \sim \begin{bmatrix} 3 & 2 & 1 & 39 \\ 0 & \frac{5}{3} & \frac{1}{3} & 8 \\ 0 & \frac{4}{3} & \frac{8}{3} & 13 \end{bmatrix} \sim \begin{bmatrix} 3 & 2 & 1 & 39 \\ 0 & \frac{5}{3} & \frac{1}{3} & 8 \\ 0 & 0 & \frac{12}{5} & \frac{33}{5} \end{bmatrix} \sim$$

$$\begin{bmatrix} 3 & 2 & 1 & 39 \\ 0 & \frac{5}{3} & \frac{1}{3} & 8 \\ 0 & 0 & \frac{12}{5} & \frac{33}{5} \end{bmatrix} \sim \begin{bmatrix} 3 & 2 & 0 & \frac{145}{4} \\ 0 & \frac{5}{3} & 0 & \frac{85}{12} \\ 0 & 0 & \frac{12}{5} & \frac{33}{5} \end{bmatrix} \sim$$

$$\begin{bmatrix} 3 & 0 & 0 & \frac{111}{4} \\ 0 & \frac{5}{3} & 0 & \frac{85}{12} \\ 0 & 0 & \frac{12}{5} & \frac{33}{5} \end{bmatrix} \sim \begin{bmatrix} 1 & 0 & 0 & \frac{37}{4} \\ 0 & 1 & 0 & \frac{17}{4} \\ 0 & 0 & 1 & \frac{11}{4} \end{bmatrix}$$

Hence, $x = \dfrac{37}{4}$, $y = \dfrac{17}{4}$, $z = \dfrac{11}{4}$.

31. We have

$$\begin{bmatrix} -1 & -1 & 0 & -1 \\ 0 & 2 & 1 & 3 \\ -2 & 1 & 3 & 7 \end{bmatrix} \sim \begin{bmatrix} -1 & -1 & 0 & -1 \\ 0 & 2 & 1 & 3 \\ 0 & 3 & 3 & 9 \end{bmatrix} \sim \begin{bmatrix} -1 & -1 & 0 & -1 \\ 0 & 2 & 1 & 3 \\ 0 & 0 & \frac{3}{2} & \frac{9}{2} \end{bmatrix} \sim$$

$$\begin{bmatrix} -1 & -1 & 0 & -1 \\ 0 & 2 & 0 & 0 \\ 0 & 0 & 1 & 3 \end{bmatrix} \sim \begin{bmatrix} -1 & 0 & 0 & -1 \\ 0 & 2 & 0 & 0 \\ 0 & 0 & 1 & 3 \end{bmatrix} \sim \begin{bmatrix} 1 & 0 & 0 & 1 \\ 0 & 1 & 0 & 0 \\ 0 & 0 & 1 & 3 \end{bmatrix}$$

So, the solution is $(1, 0, 3)$.

33. We have

$$\begin{bmatrix} 2 & -1 & 3 & -4 \\ -1 & -3 & -1 & 7 \\ 2 & -1 & 0 & 8 \end{bmatrix} \sim \begin{bmatrix} 2 & -1 & 3 & -4 \\ 0 & -\frac{7}{2} & \frac{1}{2} & 5 \\ 0 & 0 & -3 & 12 \end{bmatrix} \sim \begin{bmatrix} 2 & -1 & 3 & -4 \\ 0 & -\frac{7}{2} & \frac{1}{2} & 5 \\ 0 & 0 & 1 & -4 \end{bmatrix}$$

$$\sim \begin{bmatrix} 2 & -1 & 0 & 8 \\ 0 & -\frac{7}{2} & 0 & 7 \\ 0 & 0 & 1 & -4 \end{bmatrix} \sim \begin{bmatrix} 2 & -1 & 0 & 8 \\ 0 & 1 & 0 & -2 \\ 0 & 0 & 1 & -4 \end{bmatrix} \sim \begin{bmatrix} 2 & 0 & 0 & 6 \\ 0 & 1 & 0 & -2 \\ 0 & 0 & 1 & -4 \end{bmatrix}$$

$$\sim \begin{bmatrix} 1 & 0 & 0 & 3 \\ 0 & 1 & 0 & -2 \\ 0 & 0 & 1 & -4 \end{bmatrix}$$

So, the solution is $(3, -2, -4)$.

35. Because $\begin{bmatrix} 1 & -4 & 4 \\ 4 & -4 & 4 \end{bmatrix} \sim \begin{bmatrix} 1 & 0 & 0 \\ 0 & 1 & -1 \end{bmatrix}$, we have $\sin\theta = 0$, $\cos\theta = -1$. Hence, $\theta = (2k-1)\pi$, where k is any integer.

37. (a) $x = x_0$, $y = y_0$ is a solution so, $a_1x_0 + b_1y_0 = 0$ and $a_2x_0 + b_2y_0 = 0$. Multiply each equation by k to get $a_1(kx_0) + b_1(ky_0) = 0$ and $a_2(kx_0) + b_2(ky_0) = 0$. Therefore, $x = kx_0$, $y = ky_0$ is also a solution.

(b) Because $x = x_1$, $y = y_1$ and $x = x_2$, $y = y_2$ are solutions, we have $a_1x_1 + b_1y_1 = 0$ and $a_2x_1 + b_2y_1 = 0$ and $a_1x_2 + b_1y_2 = 0$ and $a_2x_2 + b_2y_2 = 0$. If we add first to first and second to second, we get $a_1(x_1 + x_2) + b_1(y_1 + y_2) = 0$ and $a_2(x_1 + x_2) + b_2(y_1 + y_2) = 0$. Therefore, $x = x_1 + y_1$, $y = y_1 + y_2$ is also a solution.

39. (a) If $a_1 \neq 0$, then the augmented matrix reduces to $\begin{bmatrix} a_1 & b_1 & c_1 \\ 0 & 0 & \dfrac{a_1c_2 - a_2c_1}{a_1} \end{bmatrix}$. If $a_1c_2 - a_2c_1 \neq 0$, then the system has no solutions. If $a_1c_2 - a_2c_1 = 0$, then $y = r$, $r \in \mathbf{R}$ and $x = c_1/a_1 - (b_1/a_1)r$ and the system has infinitely many solutions. If $a_1 = 0$, then $b_1a_2 = 0$. So, either $b_1 = 0$ or $a_2 = 0$. If $b_1 = 0$, then the first equation becomes $0x + 0y = c_1$. If $c_1 \neq 0$, then the system has no solutions. If $c_1 = 0$, the system reduces to the second equation only: $a_2x + b_2y = c_2$. If a_2 and b_2 are nonzero and $c_2 = 0$, then the system has no solutions. Otherwise, the system has infinitely many solutions. If now $b_1 \neq 0$, then $a_2 = 0$. So, the system reduces to $b_1y = c_1$ and $b_2y = c_2$. If $b_2 = 0$ and $c_2 = 0$, then $y = c_1/b_1$ and $x = r$, $r \in \mathbf{R}$ and the system has infinitely many solutions. If $b_2 = 0$ and $c_2 \neq 0$, the system has no solutions. If $b_2 \neq 0$ and $c_2 = 0$, there are no solutions. Lastly, suppose $b_2 \neq 0$ and $c_2 \neq 0$. Then if $c_1/b_1 = c_2/b_2$, the system has infinitely many solutions given by $x = r$, $y = c_1/b_1$. If on the other hand, $c_1/b_1 \neq c_2/b_2$, then the system has no solutions. So, in all cases the system has either no solutions, or infinitely many solutions.

(b) Because the homogeneous system has at least one solution, it should have infinitely many by part (a).

41. If x is the total number of students, then $x = x/4 + 3x/8 + 5x/16 + 2$. Solving for x yields $x = 32$.

43. There is a total of 60 people on the big boat and when they find the small boat they have food supplies for $30 - 12 = 18$ days. Let x be the number of days the food supply will last and y be the daily food consumption per person. Then $70xy$ is the total amount of food, which is also $60 \cdot 18y + 10 \cdot 4y$. Hence, $70xy = 60 \cdot 18y + 10 \cdot 4y$ or $70x = 60 \cdot 18 + 10 \cdot 4$. Therefore, $x = 16$. So, the food will last 16 days.

45. If a and b are two consecutive angles with $b > a$, then $b - a = 20$ and $2a + 2b = 360$. Hence, $b = 100$, $a = 80$, so the angles are $100°, 80°, 100°, 80°$.

47. Let b be the length of one of the equal sides and let a be the length of the third side. Suppose $b < a$. Then $a - b = 2$ and $2b + a = 16$. Hence, $b = 14/3$ and $a = 20/3$. If on the other hand, $b > a$, then $b - a = 2$ and $2b + a = 16$. In this case $b = 6$ and $a = 4$. So, we either have a triangle with sides $14/3, 14/3, 20/3$ feet, or a triangle with sides $6, 6, 4$ feet.

49. Let x be the number of squares. Then the number of black squares are $x/8 = x/7 - 5$. Therefore, $x = 280$. Hence, there are $280/20 = 14$ rows.

51. Let x mi/hr be the boats' regular speed and let y mi be the total distance. This distance is normally covered in y/x hours, but this time it took the second boat $y/x + 45/60$ hours. During the first half hour the second boat travelled $x/2$ miles, so there were $y - x/2$ miles left at speed $2x/3$. So, the total time of the second boat was also $\frac{1}{2} + \frac{y - x/2}{2x/3}$. Therefore,

$$\frac{1}{2} + \frac{y - x/2}{2x/3} = \frac{y}{x} + \frac{45}{60}$$

For the first boat we have similarly

$$\frac{x/2+4}{x}+\frac{y-x/2-4}{2x/3}=\frac{y}{x}+\frac{35}{60}$$

This system is nonlinear by if we multiply by x we get the linear system

$$\begin{aligned}-\frac{1}{4}x-2+\frac{3}{2}y&=y+\frac{7}{12}x\\ -\frac{1}{4}x+\frac{3}{2}y&=y+\frac{3}{4}x\end{aligned}$$

with solution $x=12$, $y=24$. So, the boats travelled 24 miles.

Exercises 1.2

---

1. (a) Not echelon form (b) Reduced echelon form

(c) Not echelon form (d) Echelon form but not reduced

3. (a) Echelon form but not reduced (b) Echelon form but not reduced

(c) Reduced echelon form (d) Not echelon form

5. (a) Not echelon form (b) Echelon form but not reduced

(c) Reduced echelon form (d) Reduced echelon form

7. Suppose $A \sim B$. Then there is a sequence of elementary operations, say $O_1, \ldots, O_k$, that yields B from A. Let $P_1, \ldots, P_k$ be the operations that reverse the effects of the first operations (see Exercise 6). Then the sequence $P_k, \ldots, P_1$ yields A from B. Hence, $B \sim A$.

9. Because

$$\begin{bmatrix} 1 & 1 & 1 \\ 0 & 1 & -1 \\ -1 & 0 & 0 \end{bmatrix} \sim \begin{bmatrix} 1 & 1 & 1 \\ 0 & 1 & -1 \\ 0 & 1 & 1 \end{bmatrix} \sim \begin{bmatrix} 1 & 1 & 1 \\ 0 & 1 & -1 \\ 0 & 0 & 2 \end{bmatrix}$$

$$\sim \begin{bmatrix} 1 & 1 & 1 \\ 0 & 1 & -1 \\ 0 & 0 & 1 \end{bmatrix} \sim \begin{bmatrix} 1 & 1 & 0 \\ 0 & 1 & 0 \\ 0 & 0 & 1 \end{bmatrix} \sim I$$

and

$$\begin{bmatrix} 1 & 2 & 3 \\ 2 & 2 & 3 \\ 3 & 3 & 3 \end{bmatrix} \sim \begin{bmatrix} 1 & 2 & 3 \\ 0 & -2 & -3 \\ 0 & -3 & -6 \end{bmatrix} \sim \begin{bmatrix} 1 & 2 & 3 \\ 0 & -2 & -3 \\ 0 & 0 & -\frac{3}{2} \end{bmatrix} \sim$$

$$\begin{bmatrix} 1 & 2 & 3 \\ 0 & -2 & -3 \\ 0 & 0 & 1 \end{bmatrix} \sim \begin{bmatrix} 1 & 2 & 0 \\ 0 & -2 & 0 \\ 0 & 0 & 1 \end{bmatrix} \sim \begin{bmatrix} 1 & 2 & 0 \\ 0 & 1 & 0 \\ 0 & 0 & 1 \end{bmatrix} \sim I$$

both A and B have I as a reduced row echelon form. Hence, $A \sim I$ and $B \sim I$. Therefore, $A \sim I$ and $I \sim B$, by Exercise 7. Hence, $A \sim B$ by Exercise 8.

11. Yes. The second matrix is the reduced row echelon form of the first.

13. Because the respective reduced echelon forms $\begin{bmatrix} 1 & 0 & 1 \\ 0 & 1 & 1 \end{bmatrix}$ and $\begin{bmatrix} 1 & 0 & -2 \\ 0 & 1 & 1 \end{bmatrix}$ are not equal, the matrices are not equivalent.

15. Because $\cos\theta\cos\theta - (-\sin\theta)(\sin\theta) = \cos^2\theta + \sin^2\theta = 1 \neq 0$, the matrix reduces to I by Exercise 14.

17. (a) The last two matrices of the following reduction are two echelon forms.

$$\begin{bmatrix} 1 & 1 & 1 \\ 0 & 1 & -1 \\ -1 & 0 & 0 \end{bmatrix} \sim \begin{bmatrix} 1 & 1 & 1 \\ 0 & 1 & -1 \\ 0 & 1 & 1 \end{bmatrix} \sim \begin{bmatrix} 1 & 1 & 1 \\ 0 & 1 & -1 \\ 0 & 0 & 2 \end{bmatrix} \sim \begin{bmatrix} 1 & 1 & 1 \\ 0 & 1 & -1 \\ 0 & 0 & 1 \end{bmatrix}$$

(b) The last two matrices of the following reduction are two echelon forms.

$$\begin{bmatrix} 0 & -1 & 0 & -1 \\ 1 & 0 & 1 & 1 \\ 1 & 1 & 1 & 1 \end{bmatrix} \sim \begin{bmatrix} 1 & 0 & 1 & 1 \\ 0 & -1 & 0 & -1 \\ 1 & 1 & 1 & 1 \end{bmatrix} \sim \begin{bmatrix} 1 & 0 & 1 & 1 \\ 0 & -1 & 0 & -1 \\ 0 & 1 & 0 & 0 \end{bmatrix} \sim$$

$$\sim \begin{bmatrix} 1 & 0 & 1 & 1 \\ 0 & -1 & 0 & -1 \\ 0 & 0 & 0 & -1 \end{bmatrix} \sim \begin{bmatrix} 1 & 0 & 1 & 1 \\ 0 & -1 & 0 & -1 \\ 0 & 0 & 0 & 1 \end{bmatrix}$$

19. The reduced row echelon form is the last matrix in the reduction

$$\begin{bmatrix} 0 & 0 & 0 & 0 & 0 \\ 1 & -1 & 0 & -1 & 0 \\ 0 & 0 & 1 & 0 & 1 \\ 1 & 0 & 0 & -1 & 0 \end{bmatrix} \sim \begin{bmatrix} 1 & -1 & 0 & -1 & 0 \\ 0 & 0 & 1 & 0 & 1 \\ 1 & 0 & 0 & -1 & 0 \\ 0 & 0 & 0 & 0 & 0 \end{bmatrix} \sim \begin{bmatrix} 1 & -1 & 0 & -1 & 0 \\ 0 & 0 & 1 & 0 & 1 \\ 0 & 1 & 0 & 0 & 0 \\ 0 & 0 & 0 & 0 & 0 \end{bmatrix}$$

$$\sim \begin{bmatrix} 1 & -1 & 0 & -1 & 0 \\ 0 & 1 & 0 & 0 & 0 \\ 0 & 0 & 1 & 0 & 1 \\ 0 & 0 & 0 & 0 & 0 \end{bmatrix} \sim \begin{bmatrix} 1 & 0 & 0 & -1 & 0 \\ 0 & 1 & 0 & 0 & 0 \\ 0 & 0 & 1 & 0 & 1 \\ 0 & 0 & 0 & 0 & 0 \end{bmatrix}$$

21. The reduced row echelon form is the last matrix in the reduction

$$\begin{bmatrix} -1 & -4 & 0 & -5 & 0 & -6 \\ 5 & 20 & 1 & 29 & 0 & 34 \\ 4 & 16 & 1 & 24 & 1 & 30 \end{bmatrix} \sim \begin{bmatrix} -1 & -4 & 0 & -5 & 0 & -6 \\ 0 & 0 & 1 & 4 & 0 & 4 \\ 0 & 0 & 1 & 4 & 1 & 6 \end{bmatrix} \sim$$

$$\begin{bmatrix} -1 & -4 & 0 & -5 & 0 & -6 \\ 0 & 0 & 1 & 4 & 0 & 4 \\ 0 & 0 & 0 & 0 & 1 & 2 \end{bmatrix} \sim \begin{bmatrix} 1 & 4 & 0 & 5 & 0 & 6 \\ 0 & 0 & 1 & 4 & 0 & 4 \\ 0 & 0 & 0 & 0 & 1 & 2 \end{bmatrix}$$

23. Reduction of augmented matrix of the system yields

$$\begin{bmatrix} 1 & 0 & 1 & 1 & -5 \\ 1 & 0 & -1 & 1 & -1 \\ 1 & 1 & 1 & 1 & -3 \\ 2 & 0 & 2 & 0 & -2 \end{bmatrix} \sim \begin{bmatrix} 1 & 0 & 1 & 1 & -5 \\ 0 & 0 & -2 & 0 & 4 \\ 0 & 1 & 0 & 0 & 0 \\ 0 & 0 & 0 & -2 & 8 \end{bmatrix} \sim \begin{bmatrix} 1 & 0 & 1 & 1 & -5 \\ 0 & 1 & 0 & 0 & 2 \\ 0 & 0 & -2 & 0 & 4 \\ 0 & 0 & 0 & -2 & 8 \end{bmatrix}$$

$$\sim \begin{bmatrix} 1 & 0 & 1 & 1 & -5 \\ 0 & 1 & 0 & 0 & 2 \\ 0 & 0 & 1 & 0 & -2 \\ 0 & 0 & 0 & 1 & -4 \end{bmatrix} \sim \begin{bmatrix} 1 & 0 & 1 & 0 & -1 \\ 0 & 1 & 0 & 0 & 2 \\ 0 & 0 & 1 & 0 & -2 \\ 0 & 0 & 0 & 1 & -4 \end{bmatrix} \sim \begin{bmatrix} 1 & 0 & 0 & 0 & 1 \\ 0 & 1 & 0 & 0 & 2 \\ 0 & 0 & 1 & 0 & -2 \\ 0 & 0 & 0 & 1 & -4 \end{bmatrix}$$

Therefore, $x = 1, y = 2, z = -2, w = -4$.

25. Reduction of the augmented matrix of the system yields

$$\begin{bmatrix} 1 & 0 & 4 & 5 & 0 & 0 \\ -2 & -1 & -10 & -16 & 0 & -6 \\ 2 & 6 & 20 & 46 & 1 & 33 \end{bmatrix} \sim \begin{bmatrix} 1 & 0 & 4 & 5 & 0 & 0 \\ 0 & -1 & -2 & -6 & 0 & -6 \\ 0 & 6 & 12 & 36 & 1 & 33 \end{bmatrix}$$

$$\sim \begin{bmatrix} 1 & 0 & 4 & 5 & 0 & 0 \\ 0 & -1 & -2 & -6 & 0 & -6 \\ 0 & 6 & 12 & 36 & 1 & 33 \end{bmatrix} \sim \begin{bmatrix} 1 & 0 & 4 & 5 & 0 & 0 \\ 0 & -1 & -2 & -6 & 0 & -6 \\ 0 & 0 & 0 & 0 & 1 & -3 \end{bmatrix}$$

$$\sim \begin{bmatrix} 1 & 0 & 4 & 5 & 0 & 0 \\ 0 & 1 & 2 & 6 & 0 & 6 \\ 0 & 0 & 0 & 0 & 1 & -3 \end{bmatrix}$$

Therefore, $x_1 = -4s - 5r$, $x_2 = 6 - 2s - 6r$, $x_3 = s$, $x_4 = r$, $x_5 = -3$, for any $r, s \in \mathbf{R}$.

27. Reduction of the augmented matrix of the system yields

$$\begin{bmatrix} 1 & -1 & 1 & 2 & 0 \\ 1 & 0 & 0 & 1 & -1 \\ 0 & 1 & -1 & -1 & 1 \\ 1 & 2 & 0 & 0 & -3 \end{bmatrix} \sim \begin{bmatrix} 1 & -1 & 1 & 2 & 0 \\ 0 & 1 & -1 & -1 & -1 \\ 0 & 1 & -1 & -1 & 1 \\ 0 & 3 & -1 & -2 & -3 \end{bmatrix} \sim \begin{bmatrix} 1 & -1 & 1 & 2 & 0 \\ 0 & 1 & -1 & -1 & -1 \\ 0 & 0 & 0 & 0 & 2 \\ 0 & 3 & -1 & -2 & -3 \end{bmatrix}$$

Hence, there is no solution.

29. Reduction of the augmented matrix of the system yields

$$\begin{bmatrix} 3 & 0 & 1 & 1 & -3 \\ 0 & 2 & 3 & 3 & -14 \\ 0 & 0 & 1 & -2 & 6 \\ 0 & 0 & 0 & 1 & -4 \end{bmatrix} \sim \begin{bmatrix} 3 & 0 & 1 & 0 & 1 \\ 0 & 2 & 3 & 0 & -2 \\ 0 & 0 & 1 & 0 & -2 \\ 0 & 0 & 0 & 1 & -4 \end{bmatrix} \sim$$

$$\begin{bmatrix} 3 & 0 & 0 & 0 & 3 \\ 0 & 2 & 0 & 0 & 4 \\ 0 & 0 & 1 & 0 & -2 \\ 0 & 0 & 0 & 1 & -4 \end{bmatrix} \sim \begin{bmatrix} 1 & 0 & 0 & 0 & 1 \\ 0 & 1 & 0 & 0 & 2 \\ 0 & 0 & 1 & 0 & -2 \\ 0 & 0 & 0 & 1 & -4 \end{bmatrix}$$

Hence, $x = 1$, $y = 2$, $z = -2$, $w = -4$.

31. Reduction of the augmented matrix of the system yields

$$\begin{bmatrix} 1 & 0 & 0 & 0 & -1 & -2 \\ 0 & 1 & -1 & 0 & 1 & 5 \\ 0 & -1 & 1 & 0 & -1 & -5 \\ 0 & 1 & -1 & 0 & 1 & 5 \\ 0 & -1 & 0 & -1 & 0 & -1 \end{bmatrix} \sim \begin{bmatrix} 1 & 0 & 0 & 0 & -1 & -2 \\ 0 & 1 & -1 & 0 & 1 & 5 \\ 0 & 0 & 0 & 0 & 0 & 0 \\ 0 & 0 & 0 & 0 & 0 & 0 \\ 0 & 0 & -1 & -1 & 1 & 4 \end{bmatrix} \sim$$

$$\begin{bmatrix} 1 & 0 & 0 & 0 & -1 & -2 \\ 0 & 1 & -1 & 0 & 1 & 5 \\ 0 & 0 & 1 & 1 & -1 & -4 \\ 0 & 0 & 0 & 0 & 0 & 0 \\ 0 & 0 & 0 & 0 & 0 & 0 \end{bmatrix} \sim \begin{bmatrix} 1 & 0 & 0 & 0 & -1 & -2 \\ 0 & 1 & 0 & 1 & 0 & 1 \\ 0 & 0 & 1 & 1 & -1 & -4 \\ 0 & 0 & 0 & 0 & 0 & 0 \\ 0 & 0 & 0 & 0 & 0 & 0 \end{bmatrix}$$

Hence, $x = r_1 - 2$, $y = -r_2 + 1$, $z = r_1 - r_2 - 4$, $w = r_2$, $t = r_1$.

33. Reduction of the augmented matrix of the system yields

$$\begin{bmatrix} 1 & 2 & 3 & 4 & 0 \\ 2 & 2 & 3 & 4 & 0 \\ 3 & 3 & 3 & 4 & 0 \\ 4 & 4 & 4 & 4 & 0 \end{bmatrix} \sim \begin{bmatrix} 1 & 2 & 3 & 4 & 0 \\ 0 & -2 & -3 & -4 & 0 \\ 0 & -3 & -6 & -8 & 0 \\ 0 & -4 & -8 & -12 & 0 \end{bmatrix} \sim$$

$$\begin{bmatrix} 1 & 2 & 3 & 4 & 0 \\ 0 & -2 & -3 & -4 & 0 \\ 0 & 0 & -\frac{3}{2} & -2 & 0 \\ 0 & 0 & -2 & -4 & 0 \end{bmatrix} \sim \begin{bmatrix} 1 & 2 & 3 & 4 & 0 \\ 0 & -2 & -3 & -4 & 0 \\ 0 & 0 & -\frac{3}{2} & -2 & 0 \\ 0 & 0 & 0 & 1 & 0 \end{bmatrix}$$

Back-substitution of the last system yields $x = 0, y = 0, z = 0, w = 0$.

35. Reduction of the augmented matrix yields

$$\begin{bmatrix} 1 & 1 & 0 & a \\ 1 & -1 & 0 & b \\ 1 & -1 & 1 & 0 \end{bmatrix} \sim \cdots \sim \begin{bmatrix} 1 & 0 & 0 & \frac{1}{2}a + \frac{1}{2}b \\ 0 & 1 & 0 & \frac{1}{2}a - \frac{1}{2}b \\ 0 & 0 & 1 & -b \end{bmatrix}$$

Hence, $x = \dfrac{1}{2}a + \dfrac{1}{2}b$, $y = \dfrac{1}{2}a - \dfrac{1}{2}b$, $z = -b$.

37. The matrix reduces as

$$\cdots \sim \begin{bmatrix} 1 & -1 & -2 & 2 & -7 \\ 0 & 0 & -4 & 6 & -16 \\ 0 & 0 & 0 & -3 & 6 \\ 0 & 0 & 0 & 0 & 0 \end{bmatrix} \sim \cdots \sim \begin{bmatrix} 1 & -1 & 0 & 0 & -1 \\ 0 & 0 & 1 & 0 & 1 \\ 0 & 0 & 0 & 1 & -2 \\ 0 & 0 & 0 & 0 & 0 \end{bmatrix}$$

Hence, $x_1 = r - 1$, $x_2 = r$, $x_3 = 1$, $x_4 = -2$.

39. The matrix reduces as

$$\cdots \sim \begin{bmatrix} -1 & 0 & 1 & 1 & -1 & -1 \\ 0 & 1 & 0 & -1 & -1 & 0 \\ 0 & 0 & -1 & 0 & 2 & -6 \\ 0 & 0 & 0 & 0 & 3 & -3 \\ 0 & 0 & 0 & 0 & 0 & 0 \end{bmatrix} \sim \cdots \sim \begin{bmatrix} 1 & 0 & 0 & -1 & 0 & 6 \\ 0 & 1 & 0 & -1 & 0 & -1 \\ 0 & 0 & 1 & 0 & 0 & 4 \\ 0 & 0 & 0 & 0 & 1 & -1 \\ 0 & 0 & 0 & 0 & 0 & 0 \end{bmatrix}$$

Hence, $x_1 = r + 6$, $x_2 = r - 1$, $x_3 = 4$, $x_4 = r$, $x_5 = -1$.

41. The matrix reduces as

$$\cdots \sim \begin{bmatrix} 1 & -1 & -1 & 1 & 1 & 0 \\ 0 & -2 & 0 & 2 & 0 & 0 \\ 0 & 0 & 1 & 0 & 0 & 0 \\ 0 & 0 & 0 & -1 & 1 & 0 \\ 0 & 0 & 0 & 0 & 1 & 0 \end{bmatrix} \sim \cdots \sim \begin{bmatrix} 1 & 0 & 0 & 0 & 0 & 0 \\ 0 & 1 & 0 & 0 & 0 & 0 \\ 0 & 0 & 1 & 0 & 0 & 0 \\ 0 & 0 & 0 & 1 & 0 & 0 \\ 0 & 0 & 0 & 0 & 1 & 0 \end{bmatrix}$$

Hence, $x_1 = 0$, $x_2 = 0$, $x_3 = 0$, $x_4 = 0$, $x_5 = 0$.

43. The systems are equivalent because both augmented matrices have the same reduced echelon form:

$$\begin{bmatrix} 1 & 4 & 0 & 5 & 0 & 6 \\ 0 & 0 & 1 & 4 & 0 & 4 \\ 0 & 0 & 0 & 0 & 1 & 2 \end{bmatrix}$$

45. (a) The last column is nonpivot so there are solutions. Because the third column is nonpivot there are infinitely many solutions. (b) The last column is pivot, so there are no solutions.

47. (a) If the last column is pivot, there are no solutions. Otherwise, there are infinitely many solutions. (b) No solutions (c) If the last column is pivot, then there are no solutions. Otherwise, there is exactly one solution. (d) No solutions

49. The number of pivot columns cannot exceed the number of all columns. Also, because there is at most one pivot per row, the number of pivots cannot exceed the number of rows.

51. Let A have size $m \times n$. If the first system has exactly one solution, its coefficient matrix A has n pivot columns. Hence, the second system has exactly one solution.

53. (a) The matrix reduces to $\begin{bmatrix} 2 & 3 & 4 \\ 0 & 6-\frac{3}{2}a & 8-2a \end{bmatrix}$. So, if $a = 4$, there are infinitely many solutions. If $a \neq 4$, there is exactly one solution.

(b) The matrix reduces to $\begin{bmatrix} 1 & 2 & 1 & 3 \\ 0 & 1 & -2 & 1 \\ 0 & 0 & a^2-9 & a-3 \end{bmatrix}$. So, if $a = 3$, there are infinitely many solutions. If $a = -3$, there are no solutions. If $a \neq -3, 3$ there is only one solution.

Exercises 1.3

---

1. Yes

3. No, 2 is not greater than $|1| + |1|$.

5.
$$\begin{aligned} 5x + y &= 14 \\ x - 2y &= -6 \end{aligned}$$

7. We have

k	0	1	2	3	4
x_k	1	2.6	2.1	1.94	1.99
y_k	1	3.5	4.3	4.05	3.97

So, $x_4 = 1.99$, $y_4 = 3.97$. The exact solution is $x = 2$, $y = 4$.

9. Because

k	0	1	2	3	4
x_k	0	1.2857	0.9206	0.9960	1.0011
y_k	0	4.7500	5.0714	4.9802	4.9990
z_k	0	−2.5556	−2.0278	−1.9921	−2.0022

we have $x = 1.0011$, $y = 4.9990$, $z = -2.0022$.

11. Because

k	0	1	2	3	4
x_k	0	1.0	1.96	2.072	2.0176
y_k	0	−1.8	−2.20	−2.040	−1.9952
z_k	0	3.0	3.16	3.048	2.9936

we have $x = 2.0176$, $y = -1.9952$, $z = 2.9936$.

13. Because

k	0	1	2	3	4
x_k	0	4.2500	3.9336	3.9847	3.9984
y_k	0	−0.5625	−0.9404	−0.9966	−1.0003
z_k	0	1.8281	2.0017	2.0030	2.0005

we have $x = 3.9984$, $y = -1.0003$, $z = 2.0005$.

15. (a) The first five Gauss-Seidel iterates are $(2, 0)$, $(2, -2)$, $(0, -2)$, $(0, 0)$, $(2, 0)$. Because the first and last iterates are identical, these values are repeated as k grows larger. Thus, the iteration diverges.

(b) The fourth Gauss-Seidel iterate is $(-0.5977, 0.5977)$ and the fifth one is $(-0.6006, 0.6006)$. So the iteration converges to $(-0.6, 0.6)$ to at least 2 decimal places.

17. Using partial pivoting we have

$$\begin{bmatrix} 1.2 & -4.5 & -1.23 \\ -5.5 & 1 & -15.95 \end{bmatrix} \sim \begin{bmatrix} -5.5 & 1 & -15.95 \\ 1.2 & -4.5 & -1.23 \end{bmatrix} \sim \begin{bmatrix} 1 & -0.1818 & 2.9 \\ 1.2 & -4.5 & -1.23 \end{bmatrix}$$

$$\sim \begin{bmatrix} 1 & -0.1818 & 2.9 \\ 0 & -4.2818 & -4.71 \end{bmatrix} \sim \begin{bmatrix} 1 & -0.1818 & 2.9 \\ 0 & 1 & 1.1 \end{bmatrix} \sim \begin{bmatrix} 1 & 0 & 3.1 \\ 0 & 1 & 1.1 \end{bmatrix}$$

Hence, $x = 3.1$, $y = 1.1$.

19. Using partial pivoting we have

$$\begin{bmatrix} 1.5 & 2.2 & 2.4 & 3.2 \\ 2.5 & 4.2 & 1.5 & 2.3 \\ -8.4 & 2.2 & 1.5 & -17.5 \end{bmatrix} \sim \begin{bmatrix} -8.4 & 2.2 & 1.5 & -17.5 \\ 2.5 & 4.2 & 1.5 & 2.3 \\ 1.5 & 2.2 & 2.4 & 3.2 \end{bmatrix} \sim$$

$$\begin{bmatrix} 1 & -0.2619 & -0.1786 & 2.0833 \\ 2.5 & 4.2 & 1.5 & 2.3 \\ 1.5 & 2.2 & 2.4 & 3.2 \end{bmatrix} \sim \begin{bmatrix} 1 & -0.2619 & -0.1786 & 2.0833 \\ 0 & 4.8548 & 1.9465 & -2.9083 \\ 0 & 2.5929 & 2.6679 & 0.0751 \end{bmatrix} \sim$$

$$\begin{bmatrix} 1 & -0.2619 & -0.1786 & 2.0833 \\ 0 & 1 & 0.4009 & -0.5991 \\ 0 & 2.5929 & 2.6679 & 0.0751 \end{bmatrix} \sim \begin{bmatrix} 1 & -0.2619 & -0.1786 & 2.0833 \\ 0 & 1 & 0.4009 & -0.5991 \\ 0 & 0 & 1.6284 & 1.6285 \end{bmatrix} \sim$$

$$\begin{bmatrix} 1 & -0.2619 & -0.1786 & 2.0833 \\ 0 & 1 & 0.4009 & -0.5991 \\ 0 & 0 & 1 & 1 \end{bmatrix} \sim \begin{bmatrix} 1 & -0.2619 & 0 & 2.2619 \\ 0 & 1 & 0 & -1 \\ 0 & 0 & 1 & 1 \end{bmatrix}$$

$$\sim \begin{bmatrix} 1 & 0 & 0 & 2 \\ 0 & 1 & 0 & -1 \\ 0 & 0 & 1 & 1 \end{bmatrix}$$

Hence, $x = 2$, $y = -1$, $z = 1$.

E x e r c i s e s 1.4

---

1. If there is a linear relation, say, $Ax + By = z$, then we have the system

$$\begin{aligned} 500x + 500y &= z \\ 1000x + 2000y &= z \\ 5000x + 14000y &= z \\ 10000x + 29000y &= z \end{aligned}$$

with general solution $x = \dfrac{3r}{1000}$, $y = \dfrac{-r}{1000}$, $z = r$. Therefore, $A\dfrac{3r}{1000} + B\left(\dfrac{-r}{1000}\right) = r$. So, the linear relation is $3A - B = 1000$.

3. Let x, y, z be the fixed amounts of yen, francs and marks purchased each time. Then

$$\tfrac{1}{100}x + \tfrac{1}{1.5}y + \tfrac{1}{1.2}z = 2400$$
$$\tfrac{1}{100}x + \tfrac{1}{1.2}y + \tfrac{1}{1.5}z = 2350$$
$$\tfrac{1}{125}x + \tfrac{1}{1.2}y + \tfrac{1}{1.2}z = 2390$$

Gauss elimination yields $x = 80,000$ yen, $y = 900$ francs, and $z = 1,200$ marks.

5. Let x, y, z be the temperatures in Boston, New York, and Montreal. Then

$$\begin{aligned} (x+y+z)/3 &= 30 \\ y &= (x+z)/2+9 \\ z &= (x+y)/2-9 \end{aligned}$$

Rewriting the system into standard form and using Gauss elimination we get $x = 30$, $y = 36$, and $z = 24$. Hence, the temperatures are: Boston 30°, New York 36°, and Montreal 24°.

7. Let x, y, z cm^3 be the corresponding volumes of the solutions containing A, B and C. Then

$$\begin{aligned} 1.5x + 1.8y + 3.2z &= 15.06 \\ 2x + 2.5y + 2.8z &= 17.79 \\ 1.2x + 1.5y + 3z &= 13.05 \end{aligned}$$

with solution $x = 2$, $y = 3.5$, and $z = 1.8$. Hence, the volumes of the solutions containing A, B and C are $2.0\,\mathrm{cm}^3$, $3.5\,\mathrm{cm}^3$, and $1.8\,\mathrm{cm}^3$.

9. Solving

$$\begin{array}{cccc} i_1 & -i_2 & -i_3 & = 0 \\ 2i_1 & +3i_2 & & = 6 \\ & -3i_2 & +6i_3 & = 0 \end{array}$$

we get $i_1 = \dfrac{3}{2} = 1.5$, $i_2 = 1$, $i_3 = \dfrac{1}{2} = 0.5$ Amperes.

11. By the Mean Value Property we have

$$x_1 = \tfrac{1}{4}(x_2 + 4)$$
$$x_2 = \tfrac{1}{4}(x_1 + x_3 + 4)$$
$$x_3 = \tfrac{1}{4}(x_2 + 4)$$

Hence, $x_1 = \dfrac{10}{7} = x_3$ and $x_2 = \dfrac{12}{7}$.

13. We have

$$\begin{array}{rrrrcl} 6w_1 & +6w_2 & -10w_3 & -10w_4 & = & 0 \\ 2w_1 & -8w_2 & & & = & 0 \\ & & 2w_3 & -10w_4 & = & 0 \end{array}$$

Hence, the system is balanced if $w_1 = 8r$, $w_2 = 2r$, $w_3 = 5r$, $w_4 = r$, where r is any positive real number.

15. Let $y(x) = ax^3 + bx^2 + cx + d$ be the equation of the cubic. Then

$$\begin{aligned} a\ + b + c + d &= 1 \\ -a\ + b - c + d &= 5 \\ d &= 1 \\ -8a\ + 4b - 2c + d &= 7 \end{aligned}$$

Solving yields $a = 1$, $b = 2$, $c = -3$, and $d = 1$. Hence, the equation is $y = x^3 + 2x^2 - 3x + 1$.

17. By comparing coefficients we get the system

$$\begin{aligned} a - b &= 3 - c \\ a - c &= -a \\ b + c &= -b - 2c \end{aligned}$$

Hence, $a = \frac{1}{2}, b = -\frac{3}{2}$, and $c = 1$.

19. Using common denominator and comparing numerators we get $1 = A(x-2) + B(x+1)$. So, $A + B = 0$ and $-2A + B = 1$. Therefore, $A = -\frac{1}{3}$ and $B = \frac{1}{3}$.

Exercises 2.1

1. (a) $\begin{bmatrix} 1 \\ -1 \end{bmatrix}$ (b) $\begin{bmatrix} -5 \\ -9 \end{bmatrix}$

3. (a) $\begin{bmatrix} 4 \\ -9 \\ 4 \end{bmatrix}$ (b) $\begin{bmatrix} 9 \\ -10 \\ 7 \end{bmatrix}$

5. (a) $\begin{bmatrix} -39 \\ 32 \end{bmatrix}$ (b) $\begin{bmatrix} 40 \\ 5 \end{bmatrix}$

7. The graph is

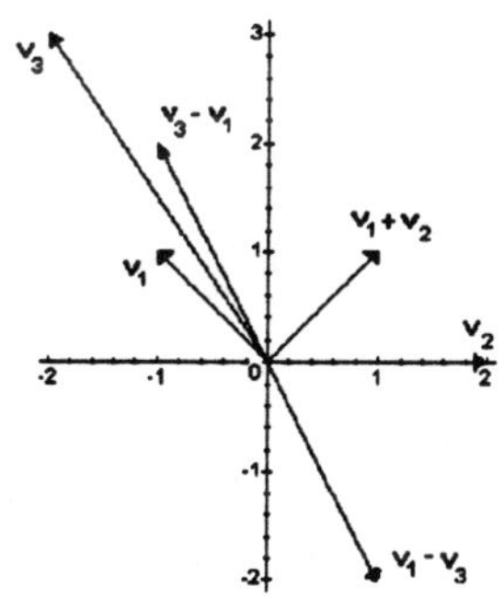

9. We have

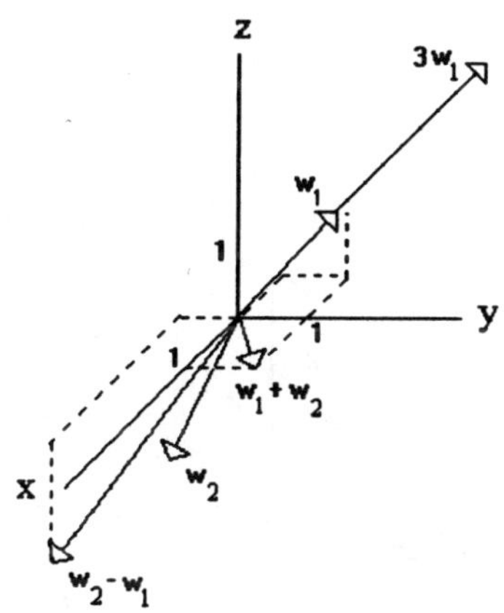

11. (a) $a = 0$, $b = 0$ (b) No values, because the last components are not equal.

13. We get $\mathbf{x} = \frac{3}{2}\mathbf{a} + 2\mathbf{b}$, because

$$2\mathbf{x} - 4\mathbf{b} = 3\mathbf{a} \Leftrightarrow 2\mathbf{x} = 3\mathbf{a} + 4\mathbf{b} \Leftrightarrow \mathbf{x} = \frac{1}{2}(3\mathbf{a} + 4\mathbf{b})$$

15. We may solve this system of vector equations just as any normal linear system with scalar variables to get $\mathbf{x} = -\mathbf{a} - 3\mathbf{b}$ and $\mathbf{y} = -2\mathbf{a} - 4\mathbf{b}$.

17. (a) $PQ = (0, -2)$ (b) $PQ = (-3, 2, 2)$

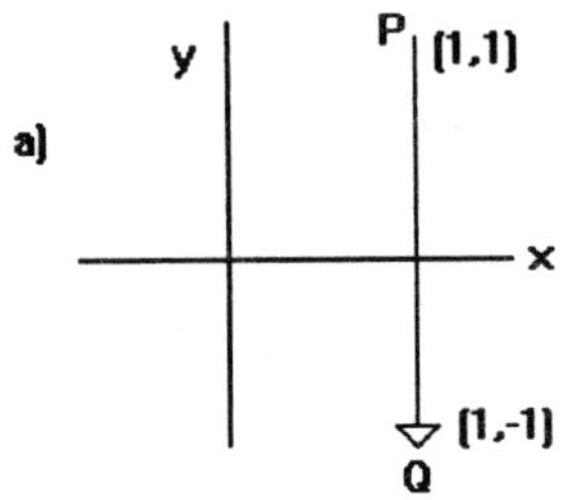

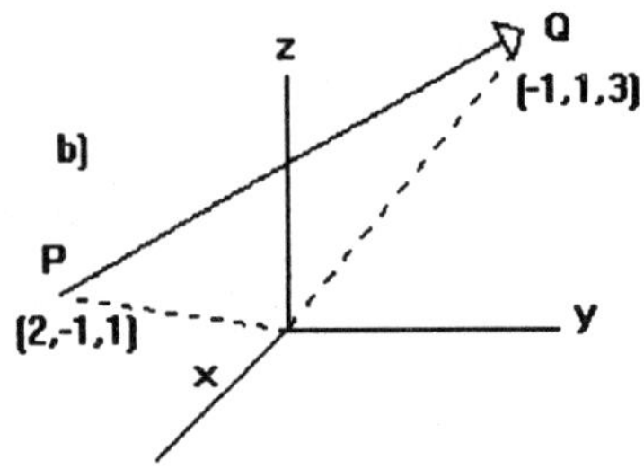

19. Yes. Because $\begin{bmatrix} a+b+c \\ a+b-2c \end{bmatrix} = (a+b)\begin{bmatrix} 1 \\ 1 \end{bmatrix} - c\begin{bmatrix} -1 \\ 2 \end{bmatrix}$.

21. No. Because $\begin{bmatrix} 1 & -1 & 1 \\ 1 & -1 & -3 \end{bmatrix} \sim \begin{bmatrix} 1 & -1 & 1 \\ 0 & 0 & -4 \end{bmatrix}$, the last column is a pivot column so the corresponding system is inconsistent.

23. No. Because $\begin{bmatrix} 1 & 0 & 1 \\ 1 & -1 & -2 \\ 1 & 1 & 0 \end{bmatrix} \sim \begin{bmatrix} 1 & 0 & 1 \\ 0 & -1 & -3 \\ 0 & 0 & -4 \end{bmatrix}$, the last column is a pivot column, so the corresponding system is inconsistent.

25. Yes. Because $\begin{bmatrix} 1 & 1 & 0 & -3 \\ 1 & 0 & -1 & 2 \\ 1 & -1 & 1 & 4 \end{bmatrix} \sim \begin{bmatrix} 1 & 1 & 0 & -3 \\ 0 & -1 & -1 & 5 \\ 0 & 0 & 3 & -3 \end{bmatrix}$, the last column is not a pivot column, so the corresponding system is consistent.

27. It is. Because $\begin{bmatrix} 1 & -1 & x \\ 3 & -2 & y \end{bmatrix} \sim \begin{bmatrix} 1 & -1 & x \\ 0 & 1 & y-3x \end{bmatrix}$, the last column is not a pivot column, so the corresponding system is consistent.

29. No, it is not. Because $\begin{bmatrix} 1 & -1 & -1 \\ 2 & 1 & 2 \\ 0 & -2 & 4 \end{bmatrix} \sim \begin{bmatrix} 1 & -1 & -1 \\ 0 & 3 & 4 \\ 0 & 0 & \frac{20}{3} \end{bmatrix}$, the corresponding system is inconsistent.

31. It is. The last column of $\begin{bmatrix} 1 & -1 & 0 & 2 \\ 0 & 1 & 0 & 1 \\ 0 & 0 & 0 & 0 \end{bmatrix}$ is not a pivot column, so the corresponding system consistent.

33. This is true, by Theorem 2. Because $\begin{bmatrix} 1 & 1 & 0 \\ 1 & 0 & -1 \\ 1 & -1 & 1 \end{bmatrix} \sim \begin{bmatrix} 1 & 1 & 0 \\ 0 & -1 & -1 \\ 0 & 0 & 3 \end{bmatrix}$ hence, each row of the first matrix has a pivot.

35. $\begin{bmatrix} 0 \\ 0 \\ 1 \end{bmatrix}$

37. $\mathbf{R}^2$

39. $\left\{ \begin{bmatrix} r \\ s \\ 0 \end{bmatrix}, r, s \in \mathbf{R} \right\}$

41. $\mathbf{R}^3$

43. Any nonzero k. Because the system with augmented matrix $\begin{bmatrix} 1 & -k & 1 & 0 \\ 0 & k & 0 & -1 \end{bmatrix}$ is consistent only if k is nonzero.

45. (a) $x \begin{bmatrix} 1 \\ -1 \\ 0 \end{bmatrix} + y \begin{bmatrix} 2 \\ 1 \\ 1 \end{bmatrix} + z \begin{bmatrix} -1 \\ 0 \\ -1 \end{bmatrix} = \begin{bmatrix} 1 \\ 0 \\ 2 \end{bmatrix}$

(b) $x \begin{bmatrix} 1 \\ -1 \\ 0 \end{bmatrix} + y \begin{bmatrix} -2 \\ 0 \\ 1 \end{bmatrix} + z \begin{bmatrix} -1 \\ 2 \\ 0 \end{bmatrix} = \begin{bmatrix} 1 \\ -1 \\ 2 \end{bmatrix}$

47. (a) No. The matrix is equivalent to $\begin{bmatrix} 1 & -1 \\ 0 & 2 \\ 0 & 0 \end{bmatrix}$ whose third row has no pivot.

(b) Yes. The matrix is equivalent to $\begin{bmatrix} -1 & 0 & 2 \\ 0 & 1 & 1 \\ 0 & 0 & 2 \end{bmatrix}$, so each row has a pivot.

49. Let u_i and v_i be the ith components of $\mathbf{u}$ and $\mathbf{v}$, respectively.

Part 2: $\mathbf{u}+\mathbf{v}$ and $\mathbf{v}+\mathbf{u}$ are both n-vectors, so they have the same size. The ith components of $\mathbf{u}+\mathbf{v}$ and $\mathbf{v}+\mathbf{u}$ are u_i+v_i and v_i+u_i, which are equal. Hence, $\mathbf{u}+\mathbf{v}=\mathbf{v}+\mathbf{u}$.

Part 3: $\mathbf{u}+\mathbf{0}$, $\mathbf{0}+\mathbf{u}$, and $\mathbf{u}$ are n-vectors. In addition their ith components are equal $u_i+0=0+u_i=u_i$. Hence, all three vectors are equal.

Part 4: $\mathbf{u}+(-\mathbf{u})$, $(-\mathbf{u})+\mathbf{u}$, and $\mathbf{0}$ are n-vectors. Their ith components are equal $u_i+(-u_i)=(-u_i)+u_i=0$. Hence, all three vectors are equal.

Part 5: $a\mathbf{u}+a\mathbf{v}$ and $a(\mathbf{u}+\mathbf{v})$ are n-vectors. Also, their ith components are equal $au_i+av_i=a(u_i+v_i)$. Hence, the vectors are equal.

51. (a) True, by Theorem 2.

(b) True, by Theorem 2.

(c) False. For example, $A=[\mathbf{e}_1\,\mathbf{e}_2]$ has two pivots, but $\mathbf{e}_3$ is not a linear combination of $\mathbf{e}_1$ and $\mathbf{e}_2$.

(d) True, by Theorem 2.

(e) False. See (c).

(f) False. The number of pivots of A must be m.

(g) False. The number of pivots of A is no more than m and n.

(h) True. Because the number of pivots of A has to be no more than m and n.

53. If the eastbound and northbound directions are along the positive x- and y-axis, respectively, then the true velocity vector is $(10, 20)$ miles per hour.

55. (a) $\begin{bmatrix} 1100 \\ 1800 \\ 2400 \end{bmatrix}$. The airline spends a total of 1100, 1800, 2400 dollars for food per trip for first, second, and third class, respectively.

(b) $\begin{bmatrix} 150 \\ 100 \\ 200 \end{bmatrix}$. The airline spends 150, 100, 200 dollars more for Plane 3 than for Plane 2 for food per trip for first, second, and third class, respectively.

(c) $\begin{bmatrix} 4500 \\ 7000 \\ 9000 \end{bmatrix}$. The airline spends 4500, 7000, and 9000 dollars for 10 trips of

Plane 3 for food for first, second, and third class, respectively.

(d) $\begin{bmatrix} 8900 \\ 14600 \\ 19300 \end{bmatrix}$. The airline spends 8900, 14600, and 9300 dollars in food for first, second, and third class, respectively, for 7 trips of the first plane, eight trips of the second, and 9 trips of the third.

Exercises 2.2

1. We have

$$\|\mathbf{u}\| = 3 \qquad \|\mathbf{v}\| = 5\sqrt{2} \qquad \|\mathbf{w}\| = 2\sqrt{5}$$

$$\|\mathbf{u}+\mathbf{v}\| = \sqrt{19} \qquad \|\mathbf{u}-\mathbf{v}\| = 3\sqrt{11} \qquad \|\mathbf{u}-\mathbf{v}+\mathbf{w}\| = \sqrt{139}$$

$$\|\mathbf{d}\| = 3 \qquad \|10\mathbf{d}\| = 30 \qquad \|\,\|\mathbf{d}\|\,\mathbf{d}\| = 9$$

3. We have

$$\mathbf{u}\cdot\mathbf{v} = -20 \qquad \mathbf{w}\cdot\mathbf{u} = 0$$

$$\mathbf{u}\cdot(\mathbf{v}+\mathbf{w}) = -20 \qquad \mathbf{v}\cdot\mathbf{u}+\mathbf{w}\cdot\mathbf{u} = -20$$

$$\mathbf{d}\cdot\mathbf{d} = 9 \qquad (\mathbf{d}\cdot\mathbf{d})\,\mathbf{d} = \left(-9,-18,9,9\sqrt{3}\right)$$

5. (a) $\dfrac{\mathbf{u}}{\|\mathbf{u}\|} = \left(-\dfrac{1}{3},\dfrac{2}{3},-\dfrac{2}{3}\right)$ (b) $\dfrac{\mathbf{v}}{\|\mathbf{v}\|} = \left(\dfrac{2\sqrt{2}}{5},-\dfrac{3\sqrt{2}}{10},\dfrac{1\sqrt{2}}{2}\right)$

(c) $\dfrac{\mathbf{w}}{\|\mathbf{w}\|} = \left(-\dfrac{2\sqrt{5}}{5},-\dfrac{\sqrt{5}}{5},0\right)$ (d) $\dfrac{\mathbf{d}}{\|\mathbf{d}\|} = \left(-\dfrac{1}{3},-\dfrac{2}{3},\dfrac{1}{3},\dfrac{\sqrt{3}}{3}\right)$

7. We need to multiply by 2 the unit vector in the opposite direction of $\mathbf{u}$. So, $2\dfrac{-\mathbf{u}}{\|\mathbf{u}\|} = \left(\dfrac{2}{3},-\dfrac{4}{3},\dfrac{4}{3}\right)$.

9. We need to multiply by 9 the unit vector in the direction of $\mathbf{u}$. So, $9\frac{\mathbf{d}}{\|\mathbf{d}\|} = \left(-3, -6, 3, 3\sqrt{3}\right)$.

11. We have $(1,1)\cdot(1,-1) = 0$, $(1,-1,-1)\cdot(0,1,-1) = 0$, $(4,-2,-1)\cdot(-2,3,4) = -18$, $(-7,-3,1,0)\cdot(0,2,6,4) = 0$. So, pairs (a), (b), and (d) are orthogonal.

13. A vector (a,b,c) orthogonal to $(-2,3,1)$ should satisfy $(-2,3,1)\cdot(a,b,c) = 0$ or $-2a+3b+c = 0$. So we may choose $(1,0,2)$. This we have to normalize to make it unit. So, $\left(\frac{\sqrt{5}}{5}, 0, \frac{2\sqrt{5}}{5}\right)$.

15. Using the formula $\cos\theta = \frac{\mathbf{u}\cdot\mathbf{v}}{\|\mathbf{u}\|\,\|\mathbf{v}\|}$ we get

(a) $\cos\theta = \frac{63}{65} \quad \Rightarrow \quad \theta \simeq 0.24871$ radians

(b) $\cos\theta = \frac{2\sqrt{3}+4}{8} \quad \Rightarrow \quad \theta \simeq 0.3681$ radians

(c) $\cos\theta = \frac{-2}{4} = -\frac{1}{2} \quad \Rightarrow \quad \theta = 2\pi/3$

17. Because $\mathbf{u_c} = \mathbf{u} - \mathbf{u_{pr}}$ we have, by Exercise 16,

(a) $\mathbf{u_c} = \left(\frac{8}{5}, \frac{16}{5}\right)$ (b) $\mathbf{u_c} = \left(\frac{33}{35}, \frac{64}{35}, \frac{9}{7}\right)$ (c) $\mathbf{u_c} = \left(-2, -1, \frac{3}{10}, \frac{1}{10}\right)$

19. By Theorem 5, $\|\mathbf{u}+\mathbf{v}\|^2 = \|\mathbf{u}\|^2 + \|\mathbf{v}\|^2 + 2\mathbf{u}\cdot\mathbf{v}$. Because $\mathbf{u}$ and $\mathbf{v}$ are orthogonal, we have $\mathbf{u}\cdot\mathbf{v} = 0$. Hence, $\|\mathbf{u}+\mathbf{v}\|^2 = \|\mathbf{u}\|^2 + \|\mathbf{v}\|^2$.

21. No. For example, $(1,1)\cdot(-1,1) = 0 = (1,1)\cdot(1,-1)$, yet $(-1,1) \neq (1,-1)$.

23. Subtracting the two equations of Theorem 5 and dividing by 4 yields the required identity.

25. The xy-plane circle of radius 1 centered at the origin.

27. If $\mathbf{v} = (v_1, v_2, v_3)$ then $\mathbf{v}\cdot\mathbf{i} = v_1$, $\mathbf{v}\cdot\mathbf{j} = v_2$, $\mathbf{v}\cdot\mathbf{k} = v_3$. Hence,

$$\cos\alpha = \frac{\mathbf{v}\cdot\mathbf{i}}{\|\mathbf{v}\|\,\|\mathbf{i}\|} = \frac{v_1}{\|\mathbf{v}\|}, \quad \cos\beta = \frac{\mathbf{v}\cdot\mathbf{j}}{\|\mathbf{v}\|\,\|\mathbf{j}\|} = \frac{v_2}{\|\mathbf{v}\|}, \quad \cos\gamma = \frac{\mathbf{v}\cdot\mathbf{k}}{\|\mathbf{v}\|\,\|\mathbf{k}\|} = \frac{v_3}{\|\mathbf{v}\|}$$

Squaring and adding these equations yields $\|(\cos\alpha, \cos\beta, \cos\gamma)\|^2 = 1$. Hence, the vector $(\cos\alpha, \cos\beta, \cos\gamma)$ is a unit vector.

29. $\|\mathbf{v}\| = \|c\mathbf{u}\| = |c|\,\|\mathbf{u}\| = c1 = c$, because $c > 0$ and $\mathbf{u}$ is unit. Hence, $\mathbf{u} = \frac{1}{c}\mathbf{v} = \frac{1}{\|\mathbf{v}\|}\mathbf{v} = (\cos\alpha, \cos\beta, \cos\gamma)$ by Equation (2.10) in the book.

31. (a) $(1,1,1) = \sqrt{3}\left(\frac{\sqrt{3}}{3}, \frac{\sqrt{3}}{3}, \frac{\sqrt{3}}{3}\right)$ with direction cosines: $\frac{\sqrt{3}}{3}, \frac{\sqrt{3}}{3}, \frac{\sqrt{3}}{3}$ and direction angles: 0.95532 radians, 0.95532 radians, and 0.95532 radians.

(b) $(-1,1,-1) = \sqrt{3}\left(-\frac{\sqrt{3}}{3}, \frac{\sqrt{3}}{3}, -\frac{\sqrt{3}}{3}\right)$ with direction cosines: $-\frac{\sqrt{3}}{3}, \frac{\sqrt{3}}{3}, -\frac{\sqrt{3}}{3}$ and direction angles: 2.1863, 0.95532, and 2.1863 radians.

(c) $(\sqrt{3},1,0) = 2\left(\frac{\sqrt{3}}{2}, \frac{1}{2}, 0\right)$ with direction cosines: $\frac{\sqrt{3}}{2}, \frac{1}{2}, 0$ and direction angles: $\frac{\pi}{6}, \frac{\pi}{3}, \frac{\pi}{2}$.

(d) $(2,-2,0) = 2\sqrt{2}\left(\frac{\sqrt{2}}{2}, -\frac{\sqrt{2}}{2}, 0\right)$ with direction cosines: $\frac{\sqrt{2}}{2}, -\frac{\sqrt{2}}{2}, 0$ and direction angles: $\frac{\pi}{4}, \frac{3\pi}{4}, \frac{\pi}{2}$.

Exercises 2.3

---

1. (a) False (b) True (c) True (d) True

3. (a) Yes. $\begin{bmatrix} -1 & 0 & 6 \\ 3 & -2 & 3 \end{bmatrix} \sim \begin{bmatrix} 1 & 0 & -6 \\ 0 & 1 & -\frac{21}{2} \end{bmatrix}$, so $\begin{bmatrix} 6 \\ 3 \end{bmatrix} = -6\begin{bmatrix} -1 \\ 3 \end{bmatrix} - \frac{21}{2}\begin{bmatrix} 0 \\ -2 \end{bmatrix}$.

(b) Yes. $\begin{bmatrix} -1 & 6 & 0 \\ 3 & 3 & -2 \end{bmatrix} \sim \begin{bmatrix} 1 & 0 & -\frac{4}{7} \\ 0 & 1 & -\frac{2}{21} \end{bmatrix}$, so $\begin{bmatrix} 0 \\ -2 \end{bmatrix} = -\frac{4}{7}\begin{bmatrix} -1 \\ 3 \end{bmatrix} - \frac{2}{21}\begin{bmatrix} 6 \\ 3 \end{bmatrix}$.

(c) Yes. $\begin{bmatrix} 0 & 6 & -1 \\ -2 & 3 & 3 \end{bmatrix} \sim \begin{bmatrix} 1 & 0 & -\frac{7}{4} \\ 0 & 1 & -\frac{1}{6} \end{bmatrix}$, so $\begin{bmatrix} -1 \\ 3 \end{bmatrix} = -\frac{7}{4}\begin{bmatrix} 0 \\ -2 \end{bmatrix} - \frac{1}{6}\begin{bmatrix} 6 \\ 3 \end{bmatrix}$.

(d) Yes. Because, $\begin{bmatrix} -3 \\ 9 \end{bmatrix} = 3\begin{bmatrix} -1 \\ 3 \end{bmatrix}$.

(e) Yes. Because, $\begin{bmatrix} -1 \\ 3 \end{bmatrix} = \frac{1}{3}\begin{bmatrix} -3 \\ 9 \end{bmatrix}$.

(f) No, **d** is not a scalar multiple of **c**.

(g) Yes. $\begin{bmatrix} 0 & 6 & -3 \\ -2 & 3 & 9 \end{bmatrix} \sim \begin{bmatrix} 1 & 0 & -\frac{21}{4} \\ 0 & 1 & -\frac{1}{2} \end{bmatrix}$, so $\begin{bmatrix} -3 \\ 9 \end{bmatrix} = -\frac{21}{4}\begin{bmatrix} 0 \\ -2 \end{bmatrix} - \frac{1}{2}\begin{bmatrix} 6 \\ 3 \end{bmatrix}$.

5. (a) Yes. Because $\begin{bmatrix} -1 & 2 & 1 & -1 \\ 3 & -3 & 0 & 0 \\ 1 & 0 & 0 & 4 \end{bmatrix} \sim \begin{bmatrix} -1 & 2 & 1 & -1 \\ 0 & 3 & 3 & -3 \\ 0 & 0 & -1 & 5 \end{bmatrix}$, the last column of the augmented matrix of the corresponding system is nonpivot; so the system is consistent.

(b) No. Because $\begin{bmatrix} -1 & 2 & -1 \\ 3 & -3 & 0 \\ 1 & 0 & 4 \end{bmatrix} \sim \begin{bmatrix} -1 & 2 & -1 \\ 0 & 3 & -3 \\ 0 & 0 & 5 \end{bmatrix}$, the last column of the augmented matrix of the corresponding system is pivot; so the system is inconsistent.

(c) No. Because $\begin{bmatrix} 2 & 1 & -1 \\ -3 & 0 & 0 \\ 0 & 0 & 4 \end{bmatrix} \sim \begin{bmatrix} 2 & 1 & -1 \\ 0 & \frac{3}{2} & -\frac{3}{2} \\ 0 & 0 & 4 \end{bmatrix}$, the last column of the augmented matrix of the corresponding system is pivot; so the system is inconsistent.

7. (a) Yes. $A = \begin{bmatrix} -1 & 2 \\ 3 & -4 \end{bmatrix} \sim \begin{bmatrix} -1 & 2 \\ 0 & 2 \end{bmatrix}$, so, A has two pivots.

(b) No. $A = \begin{bmatrix} -1 & 3 \\ 3 & -9 \end{bmatrix} \sim \begin{bmatrix} -1 & 3 \\ 0 & 0 \end{bmatrix}$, so A has only one pivot.

9. (a) Yes. $A = \begin{bmatrix} 1 & -1 & 0 \\ 7 & 3 & -1 \end{bmatrix} \sim \begin{bmatrix} 1 & -1 & 0 \\ 0 & 10 & -1 \end{bmatrix}$, so A has two pivots.

(b) No, the matrix has only two columns, so it cannot have three pivots.

11. (a) No, this matrix has only two pivots as we see by adding rows 2 and 3.
(b) No, the matrix has only three columns, so it cannot have four pivots.

13. Yes, the matrix has four pivots. It is equivalent to

$$\begin{bmatrix} 1 & 2 & -3 & 2 & -5 & 5 \\ 0 & 0 & 0 & 2 & -5 & 1 \\ 0 & 0 & 0 & 0 & 1 & 2 \\ 0 & 0 & 0 & 0 & 0 & \frac{9}{2} \end{bmatrix}$$

15. (a) False (b) True (c) False (d) True
(e) True (f) False (g) True

17. Let $V_1 = \text{Span}\{\mathbf{u}, \mathbf{v}, \mathbf{w}\}$ and let $V_2 = \text{Span}\{\mathbf{u}, \mathbf{u}+\mathbf{v}, \mathbf{u}+\mathbf{v}+\mathbf{w}\}$. Because $\mathbf{u}$, $\mathbf{v}$ and $\mathbf{w}$ are in V_1, so are $\mathbf{u}$, $\mathbf{u}+\mathbf{v}$, and $\mathbf{u}+\mathbf{v}+\mathbf{w}$ and all their linear combinations (by Theorem 10). Therefore, $V_2 \subseteq V_1$. Because, $\mathbf{v} = \mathbf{u} - (\mathbf{u} - \mathbf{v})$ and $\mathbf{w} = \mathbf{u}+\mathbf{v}+\mathbf{w} - (\mathbf{u}+\mathbf{v})$, we conclude that $\mathbf{v}$ and $\mathbf{w}$ are in V_2. Hence, all the linear combinations of $\mathbf{u}, \mathbf{v}, \mathbf{w}$ are in V_2 (by Theorem 10). Therefore, $V_1 \subseteq V_2$. Hence, $V_1 = V_2$.

19. Let $\mathbf{u}$ and $\mathbf{w}$ be in $V = \text{Span}\{\mathbf{v}_1, \dots, \mathbf{v}_n\}$. Then there are scalars $c_1, \dots, c_n$ and $d_1, \dots, d_n$ such that $\mathbf{u} = c_1\mathbf{v}_1 + \cdots + c_n\mathbf{v}_n$ and $\mathbf{v} = d_1\mathbf{v}_1 + \cdots + d_n\mathbf{v}_n$. Then $\mathbf{u}+\mathbf{v} = (c_1+d_1)\mathbf{v}_1 + \cdots + (c_n+d_n)\mathbf{v}_n$. Hence, $\mathbf{u}+\mathbf{v}$ is a linear combination of $\mathbf{v}_1, \dots, \mathbf{v}_n$, so it is in V. Also, for any scalar c the vector $c\mathbf{u} = (cc_1)\mathbf{v}_1 + \cdots + (cc_n)\mathbf{v}_n$ is a linear combination of $\mathbf{v}_1, \dots, \mathbf{v}_n$, so it too is in V. This concludes the proof of Theorem 10.

21. (a) No, we need 10 pivots.
(b) Yes. Otherwise, A would span $\mathbf{R}^{10}$ which contradicts our answer to part (a).

23. $\left\{ \begin{bmatrix} 3 \\ 0 \\ -1 \end{bmatrix}, \begin{bmatrix} -1 \\ 4 \\ 0 \end{bmatrix} \right\}$

25. That the system $[A : \mathbf{b}]$ is consistent.

27. We need to find all values of x for which $\begin{bmatrix} 1 & -1 & 0 \\ 1 & 0 & 1 \\ 0 & -1 & x \end{bmatrix} \sim \begin{bmatrix} 1 & -1 & 0 \\ 0 & 1 & 1 \\ 0 & 0 & x+1 \end{bmatrix}$ has two pivots. The latter is true if and only if $x+1 \neq 0$. So, the answer is all $x \neq -1$.

29. The corresponding spans are the shaded planes in the figures below.

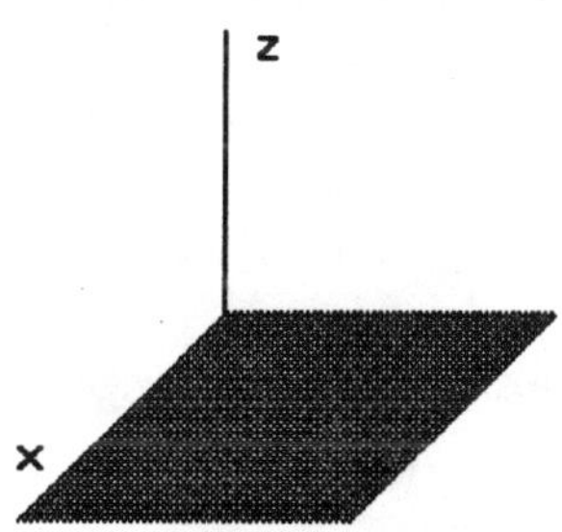

a)

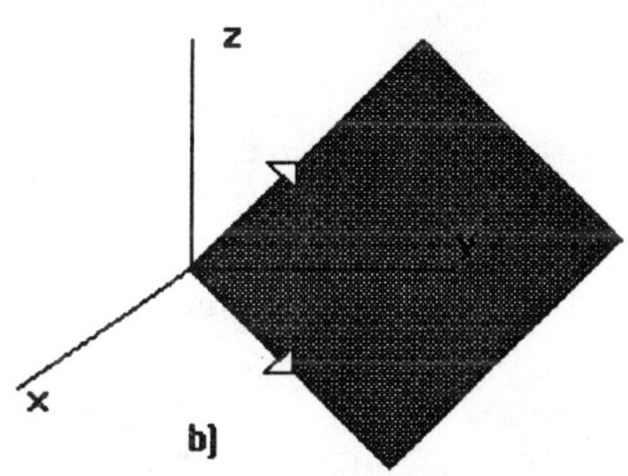

b)

Exercises 2.4

1. Linearly independent; because, $A = \begin{bmatrix} 1 & 1 & 0 \\ -2 & 1 & -1 \\ 0 & 1 & 1 \end{bmatrix} \sim \begin{bmatrix} 1 & 1 & 0 \\ 0 & 3 & -1 \\ 0 & 0 & \frac{4}{3} \end{bmatrix}$, so each column of A is a pivot column.

3. Linearly independent, because none of the vectors is a multiple of the other.

5. Linearly independent. Because, if $a \neq 0$, then $A = \begin{bmatrix} a & b & 1 \\ a & b & 0 \\ 1 & 1 & 0 \end{bmatrix} \sim \begin{bmatrix} a & b & 1 \\ 0 & \frac{a-b}{a} & -\frac{1}{a} \\ 0 & 0 & -1 \end{bmatrix}$, so each column of A is a pivot column, because $a \neq b$; if $a = 0$, then $b \neq 0$ and $A = \begin{bmatrix} 0 & b & 1 \\ 0 & b & 0 \\ 1 & 1 & 0 \end{bmatrix} \sim \begin{bmatrix} 1 & 1 & 0 \\ 0 & b & 0 \\ 0 & 0 & 1 \end{bmatrix}$, so each column of A is again a pivot column.

7. Linearly dependent, because the matrix has only one pivot column.

9. Linearly independent. The matrix has four pivot columns because it reduces to

$$\begin{bmatrix} 1 & 1 & 2 & 3 \\ 0 & 1 & 1 & 0 \\ 0 & 0 & -1 & -1 \\ 0 & 0 & 0 & 1 \end{bmatrix}$$

11. Linearly dependent; there are more vectors than there are components in each vector.

13. Linearly dependent; ten times the first vector plus twenty times the second vector yields the third vector.

15. Linearly dependent; there are more vectors than there are components in each vector.

17. Linearly independent; the vectors are pivot columns of the corresponding matrix.

19. If $a = 0$, then the vectors are $\begin{bmatrix} 0 \\ 1 \end{bmatrix}$ and $\begin{bmatrix} 2 \\ 0 \end{bmatrix}$, which are clearly linearly independent. If $a \neq 0$, then $\begin{bmatrix} a & a+2 \\ 1 & a \end{bmatrix} \sim \begin{bmatrix} a & a+2 \\ 0 & \frac{a^2-a-2}{a} \end{bmatrix}$, hence the vectors are linearly dependent if $a^2 - a - 2 = 0$; or, if $(a+1)(a-2) = 0$. So, the vectors are linearly dependent only for $a = -1, 2$.

21. Yes; there are more vectors than there are components in each vector.

23. Because $\{\mathbf{v}_1, \mathbf{v}_2, \mathbf{v}_3\}$ is linearly independent by Theorem 16, we have $c_1 = 2c_2 - c_1$, $c_2 = c_3 - c_2$, and $c_3 = c_2 - 1$. Solving the system yields $c_1 = -1$, $c_2 = -1$, $c_3 = -2$.

25. Let $c_1(\mathbf{v}_1 - \mathbf{v}_2) + c_2(\mathbf{v}_2 - \mathbf{v}_3) + c_3(\mathbf{v}_3 + \mathbf{v}_1) = \mathbf{0}$. Then $(c_1 + c_3)\mathbf{v}_1 + (c_2 - c_1)\mathbf{v}_2 + (c_3 - c_2)\mathbf{v}_3 = \mathbf{0}$. Hence, $c_1 + c_3 = 0$, $c_2 - c_1 = 0$, and $c_3 - c_2 = 0$, because $\{\mathbf{v}_1, \mathbf{v}_2, \mathbf{v}_3\}$ is linearly independent. But the system of the three equations implies that $c_1 = c_2 = c_3 = 0$. Hence, $\{\mathbf{v}_1 - \mathbf{v}_2, \mathbf{v}_2 - \mathbf{v}_3, \mathbf{v}_3 + \mathbf{v}_1\}$ is linearly independent.

27. $1 \Rightarrow 2$. Let $\mathbf{v}_1$ and $\mathbf{v}_2$ be two linearly dependent vectors. Then there are scalars c_1 and c_2 not both zero, (say, $c_1 \neq 0$) such that $c_1\mathbf{v}_1 + c_2\mathbf{v}_2 = \mathbf{0}$. Therefore, $\mathbf{v}_1 = -(c_2/c_1)\mathbf{v}_2$. Hence, $\mathbf{v}_1$ is a scalar multiple of $\mathbf{v}_2$.

$2 \Rightarrow 3$. Let one vector be a scalar multiple of the other, say, $\mathbf{v}_1 = c\mathbf{v}_2$ ($c \neq 0$, because $\mathbf{v}_1$ and $\mathbf{v}_2$ are nonzero vectors). If θ is the angle between $\mathbf{v}_1$ and $\mathbf{v}_2$, then

$$\cos\theta = \frac{\mathbf{v}_1 \cdot \mathbf{v}_2}{\|\mathbf{v}_1\|\,\|\mathbf{v}_2\|} = \frac{\mathbf{v}_1 \cdot c\mathbf{v}_1}{\|\mathbf{v}_1\|\,\|c\mathbf{v}_1\|}$$
$$= \frac{c\mathbf{v}_1 \cdot \mathbf{v}_1}{|c|\,\|\mathbf{v}_1\|^2} = \frac{c}{|c|}$$

Hence, if $c > 0$, then $\cos\theta = c/c = 1$, so, $\theta = 0$. If $c < 0$, then $\cos\theta = c/(-c) = -1$ Therefore, $\theta = \pi$. The second statement implies the third.

$3 \Rightarrow 1$. Suppose $\theta = 0$ or π. Hence, $\mathbf{v}_1 = c\mathbf{v}_2$, for some scalar c. Then $1\mathbf{v}_1 + (-c)\mathbf{v}_2 = \mathbf{0}$ shows that $\mathbf{v}_1$ and $\mathbf{v}_2$ are linearly dependent. Hence the third statement implies the first. We conclude that all three statements are equivalent.

29. Suppose $[A : \mathbf{b}]$ has two solutions, say, $x_1, \ldots, x_n$ and $x'_1, \ldots, x'_n$. If $\mathbf{a}_1, \ldots, \mathbf{a}_n$ are the columns of A then $x_1\mathbf{a}_1 + \cdots + x_n\mathbf{a}_n = \mathbf{b}$ and $x'_1\mathbf{a}_1 + \cdots + x'_n\mathbf{a}_n = \mathbf{b}$. Because the columns of A are linearly independent, $x_1 = x'_1, \ldots, x_n = x'_n$, by Theorem 16. Hence, the system $[A : \mathbf{b}]$ has at most one solution.

31. Span(S_1) and Span(S_2) are two planes through the origin. Hence, their intersection is a straight line through the origin if the planes are distinct or the common plane if the planes coincide.

Exercises 2.5

---

1. $A\mathbf{u} = \begin{bmatrix} -10 \\ -4 \\ 23 \end{bmatrix}$, whereas $A\mathbf{v}$ and $A\mathbf{w}$ are undefined.

3. If $\mathbf{x} = (x_1, x_2)$, then the system is $-3x_1 - 2x_2 = 100$, $-x_1 = 200$, $5x_1 - 3x_2 = 300$.

5. $\begin{bmatrix} 1 & -7 \\ -2 & 4 \end{bmatrix} \begin{bmatrix} x \\ y \end{bmatrix} = \begin{bmatrix} -5 \\ 0 \end{bmatrix}$

7. Let A be the matrix and $\mathbf{x}$ be the vector. Because $A\mathbf{x}$ is a 2-vector $n = 2$. A has two pivots, so the system $[A : \mathbf{b}]$ is solvable for all $\mathbf{b}$ in $\mathbf{R}^2$. Hence, the set is $\mathbf{R}^2$.

9. Let A be the matrix and $\mathbf{x}$ be the vector. Because $A\mathbf{x}$ is a 2-vector $n = 2$. A has two pivots the system $[A : \mathbf{b}]$ is solvable for all $\mathbf{b}$ in $\mathbf{R}^2$. Hence, the set is $\mathbf{R}^2$.

11. $\begin{bmatrix} 33 \\ 44 \\ 40 \end{bmatrix}$. There were 33 students in 1999, 44 in Spring 2000 and 40 in Fall 2000.

13. 6 and 18

15. Because $A\mathbf{u} = \begin{bmatrix} 0 \\ 0 \\ 0 \end{bmatrix}$, $A\mathbf{v} = \begin{bmatrix} 0 \\ -3 \\ 1 \end{bmatrix}$, and $A\mathbf{w} = \begin{bmatrix} 2 \\ 1 \\ 0 \end{bmatrix}$, only $\mathbf{u}$ is in the null space.

17. The solution set of the system is $S = \left\{ \begin{bmatrix} 10 + 3r \\ r \end{bmatrix}, r \in \mathbf{R} \right\}$. The vector $\mathbf{p} = \begin{bmatrix} 10 \\ 0 \end{bmatrix}$ is a particular solution. The null space of the coefficient matrix is the set $N = \left\{ \begin{bmatrix} 3 \\ 1 \end{bmatrix}, r \in \mathbf{R} \right\}$. Clearly $S = \mathbf{p} + N$, as claimed by Theorem 19.

19. $A(\mathbf{x}_2-\mathbf{x}_1) = A\mathbf{x}_1 - A\mathbf{x}_2 = \mathbf{b}-\mathbf{b} = \mathbf{0}$, because $\mathbf{x}_1$ and $\mathbf{x}_2$ are solutions of $A\mathbf{x} = \mathbf{b}$. Therefore, $\mathbf{x}_2 - \mathbf{x}_1$ is a solution of $A\mathbf{x} = \mathbf{0}$.

Exercises 2.6

1. We have

$$\mathbf{u} \times \mathbf{v} = (4, -3, -5)$$
$$(\mathbf{u} \times \mathbf{v}) \times \mathbf{w} = (4, -3, -5) \times \mathbf{w} = (-10, 20, -20)$$
$$\mathbf{u} \times (\mathbf{v} \times \mathbf{w}) = \mathbf{u} \times (10, -20, -20) = (-80, -40, 0)$$
$$(\mathbf{u} + \mathbf{v}) \times \mathbf{w} = (3, -1, 3) \times \mathbf{w} = (6, -12, -10)$$
$$\mathbf{u} \times \mathbf{w} + \mathbf{v} \times \mathbf{w} = (-4, 8, 10) + (10, -20, -20) = (6, -12, -10)$$
$$\mathbf{u} + (\mathbf{v} \times \mathbf{w}) = \mathbf{u} + (10, -20, -20) = (9, -18, -22)$$

3. We have

$$(\mathbf{v} \times \mathbf{w}) \cdot \mathbf{u} = \mathbf{u} \cdot (\mathbf{v} \times \mathbf{w}) = -5$$
$$\mathbf{u} \cdot (\mathbf{w} \times \mathbf{v}) = -\mathbf{u} \cdot (\mathbf{v} \times \mathbf{w}) = 5$$
$$\mathbf{v} \cdot (\mathbf{u} \times \mathbf{w}) = -\mathbf{u} \cdot (\mathbf{v} \times \mathbf{w}) = 5$$
$$\mathbf{w} \cdot (\mathbf{u} \times \mathbf{v}) = \mathbf{u} \cdot (\mathbf{v} \times \mathbf{w}) = -5$$

5. $\|\mathbf{u} \times \mathbf{v}\|^2 = \|(-29, -18, -15)\|^2 = 1390$. On the other hand, $\|\mathbf{u}\|^2 \|\mathbf{v}\|^2 - (\mathbf{u} \cdot \mathbf{v})^2 = 26(61) - 196 = 1390$.

7. Normalize $\mathbf{u} \times \mathbf{v} = (16, 12, 43)$ to get

$$\frac{1}{\sqrt{2249}}(16, 12, 43) \simeq (0.33738, 0.25304, 0.90672)$$

9. $\frac{1}{2}\|(0, -2, -2) \times (-1, 0, -2)\| = \frac{1}{2}\|(4, 2, -2)\| = \sqrt{6}$

11. $(1, -2, 3) \times (2, 0, -5) \cdot (0, 4, -1) = |(10, 11, 4) \cdot (0, 4, -1)| = |40| = 40$

13. (a) Yes, $\mathbf{u} \cdot (\mathbf{v} \times \mathbf{w}) = (-1,-1,9) \cdot (6,3,1) = 0$.

(b) No, $\mathbf{u} \cdot (\mathbf{v} \times \mathbf{w}) = (1,-1,1) \cdot (2,2,-1) = -1 \neq 0$.

15. No. For example, $(1,0,0) \times (1,0,0) = (0,0,0) = (1,0,0) \times (2,0,0)$.

17. We have

$$(\mathbf{u} \times \mathbf{v}) \cdot \mathbf{w} = (u_2v_3 - u_3v_2, u_3v_1 - u_1v_3, u_1v_2 - u_2v_1) \cdot \mathbf{w} =$$
$$w_1u_2v_3 - w_1u_3v_2 + w_2u_3v_1 - w_2u_1v_3 + w_3u_1v_2 - w_3u_2v_1$$

and

$$\mathbf{u} \cdot (\mathbf{v} \times \mathbf{w}) = \mathbf{u} \cdot (v_2w_3 - v_3w_2, v_3w_1 - v_1w_3, v_1w_2 - v_2w_1)$$

which expands to the same expression.

19. Because $V = \frac{1}{3}$(Base Area)(Height) and the base area is 1/2 the area of the parallelogram defined by $\mathbf{v}$ and $\mathbf{w}$ we have $V = \frac{1}{6}V'$, where V' is the volume of the parallelepiped defined by the tips of $\mathbf{u}$, $\mathbf{v}$, and $\mathbf{w}$. Hence, $V = \frac{1}{6}\left|\mathbf{u} \cdot (\mathbf{v} \times \mathbf{w})\right|$.

Exercises 2.7

1. Only the first two of the systems

$$\begin{array}{rcl} 5-3t &=& 5 \\ -4+2t &=& -4 \\ 2-t &=& 2 \end{array}, \quad \begin{array}{rcl} 5-3t &=& 2 \\ -4+2t &=& -2 \\ 2-t &=& 1 \end{array}, \quad \begin{array}{rcl} 5-3t &=& 1 \\ -4+2t &=& -2 \\ 2-t &=& 2 \end{array}, \quad \begin{array}{rcl} 5-3t &=& 9 \\ -4+2t &=& -12 \\ 2-t &=& -2 \end{array}$$

are consistent with respective solutions $t = 0$ and $t = 1$. We conclude that only P and Q are in line l_1.

3. For $t = 0, 1, -1$ we get respectively the points $(5,-4,2)$, $(2,-2,1)$, and $(8,-6,3)$.

5. From the corresponding direction vectors $(-3,2,-1)$, $(6,-4,2)$, $(-1,-1,1)$, $(1,2,3)$ only the first two are scalar multiples of each other. Hence, only lines l_1 and l_2 are parallel.

7. Equating the corresponding coordinates and solving the resulting system yields $t = -2$ and $s = -3$. The point of intersection is $(11, -8, 4)$.

9. $l_1 : \quad \dfrac{x-5}{-3} = \dfrac{y+4}{2} = \dfrac{z-2}{-1}, \qquad l_2 : \quad \dfrac{x-1}{6} = \dfrac{y-6}{-4} = \dfrac{z-8}{2},$

$l_3 : \quad \dfrac{x-5}{-1} = \dfrac{y+7}{-1} = \dfrac{z-11}{1}, \qquad l_4 : \quad \dfrac{x-14}{1} = \dfrac{y+2}{2} = \dfrac{z-13}{3}$

11. $x = 5 + 4t,\ y = -4 - 3t,\ z = 2 + t$

13. $x = 5 - 3t,\ y = -4 + 2t,\ z = 2 - t$

15. $\dfrac{x-3}{-4} = \dfrac{y+1}{3} = \dfrac{z+2}{7}$

17. P and Q

19. $-6(x-5) + 4(y+4) + 5(z-2) = 0$

21. $-6x + 4y + 5z + 36 = 0$

23. A normal of the plane is $PQ \times PR = (-3, 2, -1) \times (-4, 2, 0) = (2, 4, 2)$. So, we get $2(x-5) + 4(y+4) + 2(z-2) = 0$, or, $x + 2y + z + 1 = 0$.

25. The cross product of the directions of the lines yields a normal to the plane: $(-3, 2, -1) \times (1, 2, 3) = (8, 8, -8)$. Therefore, the equation is $8(x-5) + 8(y+4) - 8(z-2) = 0$, which simplifies as $x + y - z + 1 = 0$.

27. $x - 6y + 2z - 21 = 0$

29. We need the cosine of the angle of two normals of the planes. So,

$$\cos\theta = \frac{(6,1,1)\cdot(1,1,-1)}{\|(6,1,1)\|\,\|(1,1,-1)\|} = \frac{6}{\sqrt{114}}$$

31. $-x_1 + 3x_2 - 2x_3 + 8x_4 + 4x_5 + 3 = 0$

Exercises 2.8

1. We average once to get $\left(1, \frac{5}{2}, 5, \frac{9}{2}, \frac{5}{2}, 6, 5, \frac{11}{2}\right)$. Then we average for the second time to get $\left(\frac{1}{2}, \frac{7}{4}, \frac{15}{4}, \frac{19}{4}, \frac{7}{2}, \frac{17}{4}, \frac{11}{2}, \frac{21}{4}\right)$.

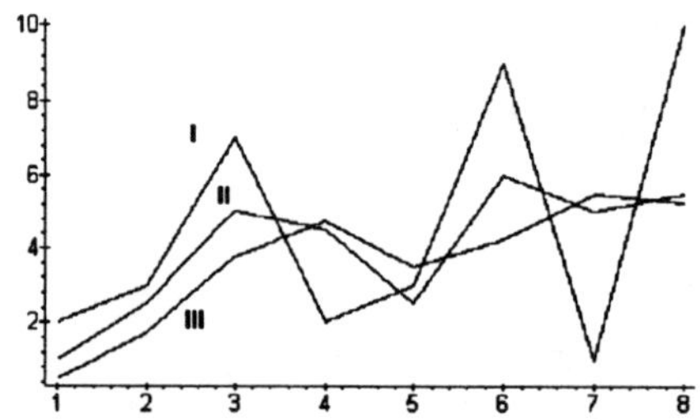

I: Original. Averaging: II: Once. III: Twice.

3. $\begin{bmatrix} Y_{k+1} \\ A_{k+1} \end{bmatrix} = \begin{bmatrix} 2 & 10 \\ \frac{4}{5} & 0 \end{bmatrix} \begin{bmatrix} Y_k \\ A_k \end{bmatrix}$. Initial condition: $\begin{bmatrix} Y_0 \\ A_0 \end{bmatrix} = \begin{bmatrix} 100 \\ 100 \end{bmatrix}$. After 3 time units: $\begin{bmatrix} Y_3 \\ A_3 \end{bmatrix} = \begin{bmatrix} 16000 \\ 2560 \end{bmatrix}$.

5. $\begin{bmatrix} Y_{k+1} \\ A_{k+1} \end{bmatrix} = \begin{bmatrix} 4 & 10 \\ \frac{1}{2} & 0 \end{bmatrix} \begin{bmatrix} Y_k \\ A_k \end{bmatrix}$. Initial condition: $\begin{bmatrix} Y_0 \\ A_0 \end{bmatrix} = \begin{bmatrix} 100 \\ 100 \end{bmatrix}$. After 3 time units: $\begin{bmatrix} Y_3 \\ A_3 \end{bmatrix} = \begin{bmatrix} 31400 \\ 3050 \end{bmatrix}$.

7. $\begin{bmatrix} Y_{k+1} \\ A_{k+1} \end{bmatrix} = \begin{bmatrix} 3 & 12 \\ \frac{1}{3} & 0 \end{bmatrix} \begin{bmatrix} Y_k \\ A_k \end{bmatrix}$. Initial condition: $\begin{bmatrix} Y_0 \\ A_0 \end{bmatrix} = \begin{bmatrix} 300 \\ 100 \end{bmatrix}$. After 3 time units: $\begin{bmatrix} Y_3 \\ A_3 \end{bmatrix} = \begin{bmatrix} 30900 \\ 2500 \end{bmatrix}$.

9. $\begin{bmatrix} A_{k+1} \\ B_{k+1} \\ C_{k+1} \end{bmatrix} = \begin{bmatrix} \frac{1}{4} & \frac{5}{2} & \frac{3}{2} \\ \frac{1}{4} & 0 & 0 \\ 0 & \frac{1}{3} & 0 \end{bmatrix} \begin{bmatrix} A_k \\ B_k \\ C_k \end{bmatrix}$. Initial condition: $\begin{bmatrix} A_0 \\ B_0 \\ C_0 \end{bmatrix} = \begin{bmatrix} 4800 \\ 4800 \\ 4800 \end{bmatrix}$. After 6 weeks: $\begin{bmatrix} A_3 \\ B_3 \\ C_3 \end{bmatrix} = \begin{bmatrix} 15975 \\ 2625 \\ 1700 \end{bmatrix}$.

11. The distance d is

$$d = \frac{|2(1) + (-2)(1) + 1(6) + (-6)|}{\sqrt{2^2 + (-2)^2 + 1^2}} = 0$$

So, the point is on the plane.

13. The distance is

$$|(4,-1,-1) \cdot ((2,-1,-1) \times (-4,4,-4))| \frac{1}{\|(2,-1,-1) \times (-4,4,-4)\|} = \frac{2\sqrt{14}}{7}$$

15. The distance is

$$|(2,-1,-1) \cdot ((4,-1,-1) \times (-2,4,-4))| \frac{1}{\|(4,-1,-1) \times (-2,4,-4)\|} = \frac{4}{73}\sqrt{146}$$

17. Because $\overrightarrow{AB}$ and $\overrightarrow{AC}$ are orthogonal we have the following implications:

$$\begin{aligned}
\overrightarrow{AB} \cdot \overrightarrow{AC} = 0 \ &\Rightarrow \ \left(\overrightarrow{AP} + \overrightarrow{PB}\right) \cdot \left(\overrightarrow{AP} + \overrightarrow{PC}\right) = 0 \\
&\Rightarrow \ \overrightarrow{AP} \cdot \overrightarrow{AP} + \overrightarrow{AP} \cdot \overrightarrow{PC} + \overrightarrow{PB} \cdot \overrightarrow{AP} + \overrightarrow{PB} \cdot \overrightarrow{PC} = 0 \\
&\Rightarrow \ \overrightarrow{AP} \cdot \overrightarrow{AP} + \overrightarrow{AP} \cdot \overrightarrow{PC} - \overrightarrow{AP} \cdot \overrightarrow{BP} + \overrightarrow{PB} \cdot \overrightarrow{PC} = 0 \\
&\Rightarrow \ \overrightarrow{AP} \cdot \overrightarrow{AP} + \overrightarrow{PB} \cdot \overrightarrow{PC} = 0 \\
&\Rightarrow \ \overrightarrow{AP} \cdot \overrightarrow{AP} = \overrightarrow{BP} \cdot \overrightarrow{PC} \\
&\Rightarrow \ \left\|\overrightarrow{AP}\right\| = \left\|\overrightarrow{BP}\right\| \\
&\Rightarrow \ AP = BP = \frac{1}{2} BC
\end{aligned}$$

19. $\frac{1}{4}((0,0,0)+(1,1,1)+(1,1,-1)+(-2,1,0)) = \left(0, \frac{3}{4}, 0\right)$

21. $\frac{1}{1+4+5+2}((-1,2,0)+4(0,5,-1)+5(1,1,-3)+2(-6,1,-3)) = \left(-\frac{2}{3}, \frac{29}{12}, -\frac{25}{12}\right)$

23. $\mathbf{R}\cdot d = (1,1,1)\cdot(1,9,-7) = 3$

25. $25\cos 45^\circ = \frac{25\sqrt{2}}{2} \simeq 17.678$ lb

Exercises 3.1

1. Matrix A has rows $\begin{bmatrix} -1 & 0 \end{bmatrix}$, $\begin{bmatrix} 2 & 3 \end{bmatrix}$, $\begin{bmatrix} -2 & 1 \end{bmatrix}$, and columns $\begin{bmatrix} -1 \\ 2 \\ -2 \end{bmatrix}$, $\begin{bmatrix} 0 \\ 3 \\ 1 \end{bmatrix}$.
A has size 3×2 with $(2,2)$ entry 3 and $(3,1)$ entry -2.
Matrix B has rows $\begin{bmatrix} -1 & 0 & -2 \end{bmatrix}$, $\begin{bmatrix} 2 & 2 & 1 \end{bmatrix}$ and columns $\begin{bmatrix} -1 \\ 2 \end{bmatrix}$, $\begin{bmatrix} 0 \\ 2 \end{bmatrix}$, $\begin{bmatrix} -2 \\ 1 \end{bmatrix}$. B has size 2×3 with $(2,2)$ entry 2 and $(2,3)$ entry 1.

3. (a) The system $x = 1$, $y - 2 = 0$, $x - y = 0$ is inconsistent.

(b) The system $x + y = 1$, $-y + z = 1$, $x + z = 3$ is inconsistent.

5. (c) and (e) are impossible due to incompatible sizes. On the other hand,

(a) $\begin{bmatrix} 1 & 0 \\ 1 & 3 \end{bmatrix}$ (b) $\begin{bmatrix} -3 & -3 & 3 \\ -3 & -3 & 3 \end{bmatrix}$

(d) $\begin{bmatrix} -9 & 9 \\ 9 & -9 \\ -9 & 9 \end{bmatrix}$ (f) $\begin{bmatrix} -12 & 26 \\ -40 & -18 \end{bmatrix}$

7. In each case we solve the equation for X and substitute the values of A, B, C to get

(a) $X = -\frac{1}{2}B - \frac{1}{2}C = \begin{bmatrix} -4 & \frac{1}{2} \\ \frac{1}{2} & -1 \end{bmatrix}$

(b) $X = \frac{1}{2}A - B + C = \begin{bmatrix} \frac{17}{2} & -4 \\ -\frac{25}{2} & 8 \end{bmatrix}$

(c) $X = 4A - 8B = \begin{bmatrix} 4 & -16 \\ -52 & 32 \end{bmatrix}$

9. Let $A = (a_{ij})$, $B = (b_{ij})$, $C = (c_{ij})$ for $i = 1, ..., m$ and $j = 1, ..., n$.

6. Clearly $(a + b)C$ and $aC + bC$ have the same size. The (i, j) entries of aC and bC are ac_{ij} and bc_{ij}. Thus, the (i, j) entry of $(a + b)C$ is $(a + b)c_{ij}$. On the other hand the (i, j) entry of $aC + bC$ is $ac_{ij} + bc_{ij}$. But $(a + b)c_{ij} = ac_{ij} + bc_{ij}$, so, $(a + b)C = aC + bC$.

7. All three matrices have the same size. Also, the (i, j) entry of aC is ac_{ij}, the (i, j) entry of bC is bc_{ij}, and the (i, j) entry of $(ab)C$ is $(ab)c_{ij}$. Therefore, the (i, j) entry of $a(bC)$ is $a(bc_{ij})$, and the (i, j) entry of $b(aC)$ is $b(ac_{ij})$. Because $(ab)c_{ij} = a(bc_{ij}) = b(ac_{ij})$, we have $(ab)C = a(bC) = b(aC)$.

8. The (i, j) entry of $1A$ is $1a_{ij} = a_{ij}$. Hence, $1A = A$.

9. The (i, j) entry of $0A$ is $0a_{ij} = 0$. Hence, $0A = 0$.

11. $\begin{bmatrix} 13 & 12 & 9 & 8 \end{bmatrix}$

13. 1

15. Because $A^2 = \begin{bmatrix} 1 & 2 \\ 0 & 1 \end{bmatrix}$, $A^4 = (A^2)^2 = \begin{bmatrix} 1 & 4 \\ 0 & 1 \end{bmatrix}$. Hence, $A^8 = (A^4)^2 = \begin{bmatrix} 1 & 8 \\ 0 & 1 \end{bmatrix}$. Guess for A^n: $A^n = \begin{bmatrix} 1 & n \\ 0 & 1 \end{bmatrix}$.

17. 2. Let A be $m \times n$ and let $B = [\mathbf{b}_1 \cdots \mathbf{b}_k]$ and $C = [\mathbf{c}_1 \cdots \mathbf{c}_k]$ be $n \times k$ matrices. Then

$$\begin{aligned} A(B + C) &= A[\mathbf{b}_1 + \mathbf{c}_1 \cdots \mathbf{b}_k + \mathbf{c}_k] \\ &= [A(\mathbf{b}_1 + \mathbf{c}_1) \cdots A(\mathbf{b}_k + \mathbf{c}_k)] \\ &= [(A\mathbf{b}_1 + A\mathbf{c}_1) \cdots (A\mathbf{b}_k + A\mathbf{c}_k)] \\ &= [A\mathbf{b}_1 \cdots A\mathbf{b}_k] + [A\mathbf{c}_1 \cdots A\mathbf{c}_k] \\ &= AB + AC \end{aligned}$$

3. Let $A = [\mathbf{a}_1 \cdots \mathbf{a}_k]$ be $n \times k$ and let B and C be $m \times n$ matrices. Then

$$\begin{aligned} (B + C)A &= (B + C)[\mathbf{a}_1 \cdots \mathbf{a}_k] \\ &= [(B + C)\mathbf{a}_1 \cdots (B + C)\mathbf{a}_k] \\ &= [(B\mathbf{a}_1 + C\mathbf{a}_1) \cdots (B\mathbf{a}_k + C\mathbf{a}_k)] \\ &= [B\mathbf{a}_1 \cdots B\mathbf{a}_k] + [C\mathbf{a}_1 \cdots C\mathbf{a}_k] \\ &= BA + CA \end{aligned}$$

4. Let B be $m \times n$ and $C = [\mathbf{c}_1 \cdots \mathbf{c}_k]$ be $n \times k$ matrices. Then

$$a(BC) = a[B\mathbf{c}_1 \cdots B\mathbf{c}_k] = [a(B\mathbf{c}_1) \cdots a(B\mathbf{c}_k)]$$

Hence, $a(BC) = [(aB)\mathbf{c}_1 \cdots (aB)\mathbf{c}_k)] = (aB)C$. Also,

$$a(BC) = [B(a\mathbf{c}_1) \cdots B(a\mathbf{c}_k)] = B(aC)$$

5. Let $A = [\mathbf{a}_1 \cdots \mathbf{a}_n]$ be $m \times n$. Then

$$I_m A = [I\mathbf{a}_1 \cdots I\mathbf{a}_k] = [\mathbf{a}_1 \cdots \mathbf{a}_k] = A$$

$$AI_n = [A\mathbf{e}_1 \cdots A\mathbf{e}_k] = [\mathbf{a}_1 \cdots \mathbf{a}_k] = A$$

6. Let $A = [\mathbf{a}_1 \cdots \mathbf{a}_n]$ be $m \times n$. Then

$$0A = [0\mathbf{a}_1 \cdots 0\mathbf{a}_k] = [0 \cdots 0] = 0$$

$$A0 = [A0 \cdots A0] = [0 \cdots 0] = 0$$

19. $A = \begin{bmatrix} 1 & 1 \\ 0 & 0 \end{bmatrix}$

21. Let $A = \begin{bmatrix} 1 & 1 \\ 0 & 0 \end{bmatrix}$, $B = \begin{bmatrix} 1 & 0 \\ 1 & 0 \end{bmatrix}$. Then $AB = \begin{bmatrix} 2 & 0 \\ 0 & 0 \end{bmatrix}$, whereas $BA = \begin{bmatrix} 1 & 1 \\ 1 & 1 \end{bmatrix}$.

23. Because $AB = BA$ we have

$$(AB)^2 = (AB)(AB) = A(BA)B = A(AB)B = (AA)(BB) = A^2B^2$$

25. $A^2 = \begin{bmatrix} -5 & 0 \\ 0 & -5 \end{bmatrix}$. $A^2A = \begin{bmatrix} -5 & 10 \\ -15 & 5 \end{bmatrix}$. Also, $A^{2+1} = A^3 = \begin{bmatrix} -5 & 10 \\ -15 & 5 \end{bmatrix}$. This verifies the first claim of the Theorem. The second claim is clear because

$$\left(A^2\right)^1 = \begin{bmatrix} -5 & 0 \\ 0 & -5 \end{bmatrix} = A^{2 \cdot 1}$$

For the last claim we have

$$(-3A)^2 = \begin{bmatrix} -3 & 6 \\ -9 & 3 \end{bmatrix}^2 = \begin{bmatrix} -45 & 0 \\ 0 & -45 \end{bmatrix}$$

and

$$(-3)^2A^2 = 9\begin{bmatrix} -5 & 0 \\ 0 & -5 \end{bmatrix} = \begin{bmatrix} -45 & 0 \\ 0 & -45 \end{bmatrix}$$

27. Because the size is limited to 2×2, we can only delete one column at a time.

$$\begin{bmatrix} 1 & 2 \\ 4 & 5 \end{bmatrix}, \quad \begin{bmatrix} 2 & 3 \\ 5 & 6 \end{bmatrix}, \quad \begin{bmatrix} 1 & 3 \\ 4 & 6 \end{bmatrix}$$

29. Direct multiplication yields

$$CD = \begin{bmatrix} -22 & 23 \\ 8 & -9 \\ 6 & -13 \\ -38 & 39 \end{bmatrix}$$

On the other hand,

$$C_{11}D_1 + C_{12}D_2 = \begin{bmatrix} -22 & 23 \\ 8 & -9 \end{bmatrix} + \begin{bmatrix} 0 & 0 \\ 0 & 0 \end{bmatrix} = \begin{bmatrix} -22 & 23 \\ 8 & -9 \end{bmatrix}$$

$$C_{21}D_1 + C_{22}D_2 = \begin{bmatrix} 9 & -13 \\ -40 & 41 \end{bmatrix} + \begin{bmatrix} -3 & 0 \\ 2 & -2 \end{bmatrix} = \begin{bmatrix} 6 & -13 \\ -38 & 39 \end{bmatrix}$$

So,

$$CD = \begin{bmatrix} C_{11}D_1 + C_{12}D_2 \\ C_{21}D_1 + C_{22}D_2 \end{bmatrix}$$

31. We must scale the matrix of the regular prices by a factor of $\frac{8}{10} = 0.8$ to get

$$\begin{bmatrix} 64 & 32 & 24 \\ 80 & 40 & 32 \\ 48 & 24 & 16 \end{bmatrix}$$

Exercises 3.2

1. $\begin{bmatrix} \frac{1}{2} & -\frac{1}{2} \\ -\frac{1}{4} & \frac{3}{4} \end{bmatrix}$, $\begin{bmatrix} \frac{1}{2} & -\frac{5}{8} \\ -\frac{1}{2} & \frac{7}{8} \end{bmatrix}$

3. We need the inverse of the given matrix: $\begin{bmatrix} 4 & -\frac{3}{2} \\ -1 & \frac{1}{2} \end{bmatrix}$.

5. A is invertible because $a^2 + b^2 \neq 0$. So, $A^{-1} = \dfrac{1}{a^2+b^2}\begin{bmatrix} a & -b \\ b & a \end{bmatrix} = \begin{bmatrix} a & -b \\ b & a \end{bmatrix}$.

7. By Theorem 7,

$$(2A)^3 = 2^3A^3 = 8\begin{bmatrix} 1 & 1 \\ -5 & -2 \end{bmatrix} = \begin{bmatrix} 8 & 8 \\ -40 & -16 \end{bmatrix}$$

Hence,

$$(2A)^{-3} = \begin{bmatrix} 8 & 8 \\ -40 & -16 \end{bmatrix}^{-1} = \begin{bmatrix} -\frac{1}{12} & -\frac{1}{24} \\ \frac{5}{24} & \frac{1}{24} \end{bmatrix}$$

9. $\begin{bmatrix} -1 & 0 & 0 \\ 0 & 1 & 0 \\ 0 & 0 & -1 \end{bmatrix}$, $\begin{bmatrix} \frac{1}{5} & 0 & 0 \\ 0 & \frac{1}{5} & 0 \\ 0 & 0 & \frac{1}{5} \end{bmatrix}$

11. Let B be an inverse. Then $AB = I$. Because the first row of A is zero, the $(1,1)$ entry of AB is 0. This contradicts the fact that the $(1,1)$ entry of I is 1. Therefore, the matrix has no inverse.

13. (a) We have

$$\begin{bmatrix} -1 & 1 & 0 & 1 & 0 & 0 \\ 0 & -1 & 0 & 0 & 1 & 0 \\ 0 & 0 & 1 & 0 & 0 & 1 \end{bmatrix} \sim \begin{bmatrix} -1 & 0 & 0 & 1 & 1 & 0 \\ 0 & 1 & 0 & 0 & -1 & 0 \\ 0 & 0 & 1 & 0 & 0 & 1 \end{bmatrix} \sim \begin{bmatrix} 1 & 0 & 0 & -1 & -1 & 0 \\ 0 & 1 & 0 & 0 & -1 & 0 \\ 0 & 0 & 1 & 0 & 0 & 1 \end{bmatrix}$$

So the inverse is $\begin{bmatrix} -1 & -1 & 0 \\ 0 & -1 & 0 \\ 0 & 0 & 1 \end{bmatrix}$.

(b) We have

$$\begin{bmatrix} -2 & 0 & -1 & 1 & 0 & 0 \\ 0 & -1 & -2 & 0 & 1 & 0 \\ 0 & 0 & -2 & 0 & 0 & 1 \end{bmatrix} \sim \begin{bmatrix} -2 & 0 & 0 & 1 & 0 & -\frac{1}{2} \\ 0 & -1 & 0 & 0 & 1 & -1 \\ 0 & 0 & 1 & 0 & 0 & -\frac{1}{2} \end{bmatrix} \sim$$

$$\begin{bmatrix} -2 & 0 & 0 & 1 & 0 & -\frac{1}{2} \\ 0 & 1 & 0 & 0 & -1 & 1 \\ 0 & 0 & 1 & 0 & 0 & -\frac{1}{2} \end{bmatrix} \sim \begin{bmatrix} 1 & 0 & 0 & -\frac{1}{2} & 0 & \frac{1}{4} \\ 0 & 1 & 0 & 0 & -1 & 1 \\ 0 & 0 & 1 & 0 & 0 & -\frac{1}{2} \end{bmatrix}$$

So the inverse is $\begin{bmatrix} -\frac{1}{2} & 0 & \frac{1}{4} \\ 0 & -1 & 1 \\ 0 & 0 & -\frac{1}{2} \end{bmatrix}$.

(c) We have

$$\begin{bmatrix} 1 & 2 & -1 & 1 & 0 & 0 \\ 1 & -2 & -1 & 0 & 1 & 0 \\ 1 & 6 & -1 & 0 & 0 & 1 \end{bmatrix} \sim \begin{bmatrix} 1 & 2 & -1 & 1 & 0 & 0 \\ 0 & -4 & 0 & -1 & 1 & 0 \\ 0 & 4 & 0 & -1 & 0 & 1 \end{bmatrix} \sim$$

$$\begin{bmatrix} 1 & 2 & -1 & 1 & 0 & 0 \\ 0 & -4 & 0 & -1 & 1 & 0 \\ 0 & 0 & 0 & -2 & 1 & 1 \end{bmatrix}$$

Therefore, the matrix is noninvertible.

(d) We have

$$\begin{bmatrix} -1 & 1 & 0 & 1 & 0 & 0 \\ 1 & 0 & -1 & 0 & 1 & 0 \\ 1 & 1 & -1 & 0 & 0 & 1 \end{bmatrix} \sim \begin{bmatrix} -1 & 1 & 0 & 1 & 0 & 0 \\ 0 & 1 & -1 & 1 & 1 & 0 \\ 0 & 2 & -1 & 1 & 0 & 1 \end{bmatrix} \sim$$
$$\begin{bmatrix} -1 & 1 & 0 & 1 & 0 & 0 \\ 0 & 1 & -1 & 1 & 1 & 0 \\ 0 & 0 & 1 & -1 & -2 & 1 \end{bmatrix} \sim \begin{bmatrix} -1 & 1 & 0 & 1 & 0 & 0 \\ 0 & 1 & 0 & 0 & -1 & 1 \\ 0 & 0 & 1 & -1 & -2 & 1 \end{bmatrix} \sim$$
$$\begin{bmatrix} -1 & 0 & 0 & 1 & 1 & -1 \\ 0 & 1 & 0 & 0 & -1 & 1 \\ 0 & 0 & 1 & -1 & -2 & 1 \end{bmatrix} \sim \begin{bmatrix} 1 & 0 & 0 & -1 & -1 & 1 \\ 0 & 1 & 0 & 0 & -1 & 1 \\ 0 & 0 & 1 & -1 & -2 & 1 \end{bmatrix}$$

Hence, the inverse is $\begin{bmatrix} -1 & -1 & 1 \\ 0 & -1 & 1 \\ -1 & -2 & 1 \end{bmatrix}$.

15. We have

$$\begin{bmatrix} -1 & 1 & 1 & -1 & 1 & 0 & 0 & 0 \\ -1 & 0 & 1 & 0 & 0 & 1 & 0 & 0 \\ 0 & 1 & -1 & 1 & 0 & 0 & 1 & 0 \\ 0 & 0 & 1 & -1 & 0 & 0 & 0 & 1 \end{bmatrix} \sim \begin{bmatrix} -1 & 1 & 1 & -1 & 1 & 0 & 0 & 0 \\ 0 & -1 & 0 & 1 & -1 & 1 & 0 & 0 \\ 0 & 1 & -1 & 1 & 0 & 0 & 1 & 0 \\ 0 & 0 & 1 & -1 & 0 & 0 & 0 & 1 \end{bmatrix} \sim$$
$$\begin{bmatrix} -1 & 1 & 1 & -1 & 1 & 0 & 0 & 0 \\ 0 & -1 & 0 & 1 & -1 & 1 & 0 & 0 \\ 0 & 0 & -1 & 2 & -1 & 1 & 1 & 0 \\ 0 & 0 & 1 & -1 & 0 & 0 & 0 & 1 \end{bmatrix} \sim \begin{bmatrix} -1 & 1 & 1 & -1 & 1 & 0 & 0 & 0 \\ 0 & -1 & 0 & 1 & -1 & 1 & 0 & 0 \\ 0 & 0 & -1 & 2 & -1 & 1 & 1 & 0 \\ 0 & 0 & 0 & 1 & -1 & 1 & 1 & 1 \end{bmatrix} \sim$$
$$\begin{bmatrix} -1 & 1 & 1 & 0 & 0 & 1 & 1 & 1 \\ 0 & -1 & 0 & 0 & 0 & 0 & -1 & -1 \\ 0 & 0 & -1 & 0 & 1 & -1 & -1 & -2 \\ 0 & 0 & 0 & 1 & -1 & 1 & 1 & 1 \end{bmatrix} \sim \begin{bmatrix} -1 & 1 & 0 & 0 & 1 & 0 & 0 & -1 \\ 0 & -1 & 0 & 0 & 0 & 0 & -1 & -1 \\ 0 & 0 & 1 & 0 & -1 & 1 & 1 & 2 \\ 0 & 0 & 0 & 1 & -1 & 1 & 1 & 1 \end{bmatrix}$$
$$\sim \begin{bmatrix} -1 & 0 & 0 & 0 & 1 & 0 & -1 & -2 \\ 0 & 1 & 0 & 0 & 0 & 0 & 1 & 1 \\ 0 & 0 & 1 & 0 & -1 & 1 & 1 & 2 \\ 0 & 0 & 0 & 1 & -1 & 1 & 1 & 1 \end{bmatrix} \sim \begin{bmatrix} 1 & 0 & 0 & 0 & -1 & 0 & 1 & 2 \\ 0 & 1 & 0 & 0 & 0 & 0 & 1 & 1 \\ 0 & 0 & 1 & 0 & -1 & 1 & 1 & 2 \\ 0 & 0 & 0 & 1 & -1 & 1 & 1 & 1 \end{bmatrix}$$

So the inverse is

$$\begin{bmatrix} -1 & 0 & 1 & 2 \\ 0 & 0 & 1 & 1 \\ -1 & 1 & 1 & 2 \\ -1 & 1 & 1 & 1 \end{bmatrix}$$

17. We have

(a) $\begin{bmatrix} 1 & 1 & -1 \\ 1 & 0 & -1 \\ -1 & 1 & 0 \end{bmatrix}\begin{bmatrix} 1 \\ 2 \\ 3 \end{bmatrix} = \begin{bmatrix} -4 \\ -1 \\ -6 \end{bmatrix}$ (b) $\begin{bmatrix} 1 & 1 & -1 \\ 1 & 0 & -1 \\ -1 & 1 & 0 \end{bmatrix}\begin{bmatrix} 2 \\ 4 \\ -1 \end{bmatrix} = \begin{bmatrix} 7 \\ 3 \\ 2 \end{bmatrix}$

19. $A = (A^{-1})^{-1} = \begin{bmatrix} \frac{1}{2} & \frac{1}{2} & -\frac{1}{2} \\ \frac{1}{2} & -\frac{1}{2} & \frac{1}{2} \\ 0 & 0 & 1 \end{bmatrix}$

21. We have

$$A^{-1} = \begin{bmatrix} 0 & 0 & \frac{1}{c_3} \\ 0 & \frac{1}{c_2} & 0 \\ \frac{1}{c_1} & 0 & 0 \end{bmatrix} \qquad B^{-1} = \begin{bmatrix} 0 & 0 & 0 & \frac{1}{c_4} \\ 0 & 0 & \frac{1}{c_3} & 0 \\ 0 & \frac{1}{c_2} & 0 & 0 \\ \frac{1}{c_1} & 0 & 0 & 0 \end{bmatrix}$$

23. $A^{-1} = \begin{bmatrix} 1 & 0 & 0 \\ 0 & -1 & 1 \\ 0 & 0 & 1 \end{bmatrix}$, $A^{-2} = I$. Therefore, $A^{-2k} = I$ and $A^{-(2k-1)} = A^{-1}$, for any positive integer k. In particular, $A^{-24} = I$ and $A^{-25} = A^{-1}$.

25. $A = \begin{bmatrix} 1 & 0 \\ 0 & 1 \end{bmatrix}$, $B = \begin{bmatrix} 0 & 2 \\ 1 & 0 \end{bmatrix}$

27. Let the row of zeros be the ith row and let B be an inverse. Then $AB = I$. Because the ith row of A is zero, the (i, i) entry of AB is 0. This cannot be the case, because the (i, i) entry of I is 1. Therefore, the matrix has no inverse.

29. Let A be an invertible diagonal matrix. We shall show that all the diagonal entries a_{ii}, $i = 1, \ldots, n$ of A are nonzero and that A^{-1} is diagonal with diagonal entries a_{ii}^{-1}. Let A^{-1} have entries b_{ij}. Because $AA^{-1} = I$ the (i, j) entry $\Sigma_k a_{ik} b_{kj}$ of the product AA^{-1} is 1 if $i = j$, and 0 if $i \neq j$. If $i = j$ this entry reduces to $a_{ii}b_{ii} = 1$, because $a_{ik} = 0$ for $i \neq k$. Hence, a_{ii} is not zero and $b_{ii} = a_{ii}^{-1}$. Moreover, if $i \neq j$ then the entry reduces to $a_{ii}b_{ij} = 0$. But because $a_{ii} \neq 0$ we conclude that $b_{ij} = 0$, for $i \neq j$. Therefore, A^{-1} is diagonal with diagonal entries a_{ii}^{-1}. Conversely, let A be diagonal with entirely nonzero diagonal entries a_{ii}. Consider the diagonal matrix B with diagonal entries a_{ii}^{-1}. The product AB is clearly the identity matrix I. Hence, A is invertible with inverse B.

31. If on the contrary A^{-1} existed then

$$AB = \mathbf{0} \Rightarrow A^{-1}AB = A^{-1}\mathbf{0} \Rightarrow B = \mathbf{0}$$

which is a contradiction because $B \neq \mathbf{0}$.

33. The claim follows from

$$(I - A)(A^2 + A + I) = IA^2 + IA + I^2 - A^3 - A^2 - AI = I - A^3 = I$$
$$(A^2 + A + I)(I - A) = A^2I - A^3 + AI - A^2 + I^2 - IA = I - A^3 = I$$

35. Let $D = ac - bd$. Assume $D \neq 0$ and let $A' = \frac{1}{D}\begin{bmatrix} d & -b \\ -c & a \end{bmatrix}$. Then

$$AA' = \begin{bmatrix} \frac{ad - bc}{D} & 0 \\ 0 & \frac{ad - bc}{D} \end{bmatrix} = I = A'A$$

Hence, A is invertible and A' is its inverse.
Conversely, assume A is invertible. We shall show that $D \neq 0$. Because A is invertible the system $A\mathbf{x} = \mathbf{e}_1$ is consistent and if $a \neq 0$ it reduces to $\begin{bmatrix} a & b & 0 \\ 0 & \frac{D}{a} & 1 \end{bmatrix}$. Hence D is nonzero, or else the last column would be a pivot column. Suppose now $a = 0$. Then because the system $A\mathbf{x} = \mathbf{e}_2$ is consistent and from the row reduction $\begin{bmatrix} c & d & 0 \\ 0 & b & 1 \end{bmatrix}$ we conclude that $b \neq 0$. To show that $D \neq 0$ it suffices to show that c is nonzero (because $a = 0$). If c were zero then the consistent system $A\mathbf{x} = \mathbf{e}_1$ would reduce to $\begin{bmatrix} 0 & b & 1 \\ 0 & 0 & -\frac{d}{b} \end{bmatrix}$. Hence, $d = 0$. This is impossible because the system $A\mathbf{x} = \mathbf{e}_2$ (with augmented matrix $\begin{bmatrix} 0 & b & 0 \\ 0 & 0 & 1 \end{bmatrix}$) is inconsistent. So c is nonzero and so is D.

37. (a) If $n \geq 0$, we have

$$\begin{aligned} A^n \left(A^{-1}\right)^n &= (A \cdots A)(A^{-1} \cdots A^{-1}) \\ &= A \cdots (AA^{-1}) \cdots A^{-1} \\ &= A \cdots I \cdots A^{-1} \\ &\vdots \\ &= I \end{aligned}$$

Similarly, $\left(A^{-1}\right)^n A^n = I$. Hence, A^n is invertible with inverse $\left(A^{-1}\right)^n$. If $n < 0$, we repeat the argument with $-n$ in place of n.

(b) The case where both m and n are non negative was covered in Theorem 3. If $n < 0$ and $m \geq 0$ and $m \geq -n$, then

$$A^n A^m = \left(A^{-1}\right)^{-n} A^m = \underbrace{A^{-1} \cdots A^{-1}}_{-n} \underbrace{A \cdots A}_{m} = \underbrace{A \cdots A}_{m+n}$$

The other cases where at least one of m and n are negative are proved similarly.

(c) The case where both m and n are non negative was covered in Theorem 3. If $n < 0$ and $m > 0$, then by Theorem 3 and by part (a)

$$(A^n)^m = \left(\left(A^{-1}\right)^{-n}\right)^m = \left(A^{-1}\right)^{-nm} = A^{mn}$$

The other cases are proved similarly.

(d) It suffices to consider $n < 0$. By Theorems 6 and 3

$$\begin{aligned} (cA)^n &= \left((cA)^{-1}\right)^{-n} = \left(\frac{1}{c}A^{-1}\right)^{-n} \\ &= \left(\frac{1}{c}\right)^{-n} \left(A^{-1}\right)^{-n} = c^n A^n \end{aligned}$$

39. (a) Part 1. A^T has columns the rows of A, hence $\left(A^T\right)^T$ has rows the rows of A. So, $\left(A^T\right)^T = A$. Part 2. The rows $(A+B)^T$ are columns of $A+B$ which in turn are the rows of $A^T + B^T$. So, $(A+B)^T = A^T + B^T$. Part 3. The rows of $(cA)^T$

are the columns of cA which in turn are the rows of cA^T. So, $(cA)^T = cA^T$. Part 5. Suppose A is invertible. By part 4 we have

$$A^T\left(A^{-1}\right)^T = \left(A^{-1}A\right)^T = I^T = I$$

Similarly, $\left(A^{-1}\right)^T A^T = I$. Hence, A^T is invertible and $\left(A^{-1}\right)^T$ is its inverse.

(b) First we note that if $\mathbf{v}$ is $1 \times m$ and A is $m \times n$ then by the matrix×vector case proved in part 4 we have $\mathbf{v}A = \left(\mathbf{v}^T\right)^T \left(A^T\right)^T = \left(A^T\mathbf{v}^T\right)^T$. Hence, $(\mathbf{v}A)^T = A^T\mathbf{v}^T$. Now let B be any $n \times k$ matrix. Then

$$\begin{aligned} j\text{th column of } (AB)^T &= (j\text{th row of } AB)^T \\ &= (j\text{th row of } A \text{ times } B)^T \\ &= B^T \text{ times the } (j\text{th row of } A)^T \\ &= B^T \text{ times the } j\text{th column of } A^T \\ &= j\text{th column of } B^T A^T \end{aligned}$$

The third equality holds by the $1 \times m$ by $m \times n$ case. This completes the proof of part 4.

41. AA^T is symmetric because

$$\left(AA^T\right)^T = \left(A^T\right)^T A^T = AA^T$$

Exercises 3.3

---

1. A, C, D are elementary. Corresponding operations: $R_1+R_2 \to R_1$, $R_1-2R_3 \to R_1$, and $R_2 \leftrightarrow R_4$.

3. $R_1 \leftrightarrow R_2$, $2R_2 \to R_2$, $R_1 - 5R_3 \to R_1$, and $-R_1 + R_3 \to R_3$.

5. (a) $R_1 \leftrightarrow R_3$ (b) $R_1 - 5R_3 \to R_1$ (c) $2R_4 \to R_4$

(d) $-10R_1 + R_3 \to R_3$

7. We have

$$A = \begin{bmatrix} 1 & 0 \\ -\frac{1}{2} & 1 \end{bmatrix} \begin{bmatrix} 1 & 0 \\ 0 & \frac{1}{2} \end{bmatrix} \begin{bmatrix} 1 & 1 \\ 0 & 1 \end{bmatrix} \begin{bmatrix} 2 & 0 \\ 0 & 1 \end{bmatrix}$$

$$A^{-1} = \begin{bmatrix} \frac{1}{2} & 0 \\ 0 & 1 \end{bmatrix} \begin{bmatrix} 1 & -1 \\ 0 & 1 \end{bmatrix} \begin{bmatrix} 1 & 0 \\ 0 & 2 \end{bmatrix} \begin{bmatrix} 1 & 0 \\ \frac{1}{2} & 1 \end{bmatrix}$$

9. We have

$$A = \begin{bmatrix} 1 & -2 \\ 0 & 1 \end{bmatrix} \begin{bmatrix} 1 & 0 \\ 0 & 3 \end{bmatrix} \begin{bmatrix} 3 & 0 \\ 0 & 1 \end{bmatrix}$$

$$B = \begin{bmatrix} 2 & 0 \\ 0 & 1 \end{bmatrix} \begin{bmatrix} 1 & 0 \\ 1 & 1 \end{bmatrix}$$

$$C = \begin{bmatrix} 1 & 0 & 0 \\ 0 & 1 & 0 \\ 0 & 1 & 1 \end{bmatrix} \begin{bmatrix} 1 & 0 & 0 \\ 0 & 1 & 0 \\ 1 & 0 & 1 \end{bmatrix}$$

11. The matrix is singular because its reduced row echelon form

$$\begin{bmatrix} 1 & 0 & 1 \\ 0 & 1 & 0 \\ 0 & 0 & 0 \end{bmatrix}$$

has a row of zeros.

13. This is true by Theorem 15, because the coefficient matrix is invertible. More precisely

$$\begin{bmatrix} 1 & -1 \\ 1 & 2 \end{bmatrix}^{-1} = \begin{bmatrix} \frac{2}{3} & \frac{1}{3} \\ -\frac{1}{3} & \frac{1}{3} \end{bmatrix}$$

15. We have

$$A = \begin{bmatrix} 0 & 0 & 1 \\ 0 & 1 & 0 \\ 1 & 0 & 0 \end{bmatrix} \begin{bmatrix} c_3 & 0 & 0 \\ 0 & 1 & 0 \\ 0 & 0 & 1 \end{bmatrix} \begin{bmatrix} 1 & 0 & 0 \\ 0 & c_2 & 0 \\ 0 & 0 & 1 \end{bmatrix} \begin{bmatrix} 1 & 0 & 0 \\ 0 & 1 & 0 \\ 0 & 0 & c_1 \end{bmatrix}$$

and

$$A^{-1} = \begin{bmatrix} 0 & 0 & c_3^{-1} \\ 0 & c_2^{-1} & 0 \\ c_1^{-1} & 0 & 0 \end{bmatrix}$$

$$= \begin{bmatrix} 1 & 0 & 0 \\ 0 & 1 & 0 \\ 0 & 0 & c_1^{-1} \end{bmatrix} \begin{bmatrix} 1 & 0 & 0 \\ 0 & c_2^{-1} & 0 \\ 0 & 0 & 1 \end{bmatrix} \begin{bmatrix} c_3^{-1} & 0 & 0 \\ 0 & 1 & 0 \\ 0 & 0 & 1 \end{bmatrix} \begin{bmatrix} 0 & 0 & 1 \\ 0 & 1 & 0 \\ 1 & 0 & 0 \end{bmatrix}$$

17. Let A be the given matrix. 1. A is invertible. In fact, $A^{-1} = \begin{bmatrix} 3 & 0 & -2 \\ 0 & 1 & 0 \\ 2 & 0 & -1 \end{bmatrix}$.

2. $A \sim I$ because I is the reduced row echelon form of A.

3.

$$A = \begin{bmatrix} 1 & 0 & 0 \\ 0 & 1 & 0 \\ 2 & 0 & 1 \end{bmatrix} \begin{bmatrix} 1 & 0 & 0 \\ 0 & 1 & 0 \\ 0 & 0 & -1 \end{bmatrix} \begin{bmatrix} 1 & 0 & 2 \\ 0 & 1 & 0 \\ 0 & 0 & 1 \end{bmatrix} \begin{bmatrix} -1 & 0 & 0 \\ 0 & 1 & 0 \\ 0 & 0 & 1 \end{bmatrix}$$

4. We may take B to be A^{-1}.

5. We may take C to be A^{-1}.

6 and 7. Because $A \sim \begin{bmatrix} -1 & 0 & 2 \\ 0 & 1 & 0 \\ 0 & 0 & -1 \end{bmatrix}$ we see that A has 3 pivots, hence each column is a pivot column and each row has a pivot.

8. The columns are linearly independent because each column is a pivot column by part 6.

9. The rows are linearly independent because $A^T \sim \begin{bmatrix} -1 & 0 & -2 \\ 0 & 1 & 0 \\ 0 & 0 & -1 \end{bmatrix}$ has 3 pivots.

19. (a) False. For example, $\begin{bmatrix} 1 & 0 \\ 0 & 1 \\ 0 & 0 \end{bmatrix}$ is noninvertible with linearly independent columns.

(b) True

(c) False. $\begin{bmatrix} 1 & 0 & 0 \\ 0 & 1 & 0 \end{bmatrix} \begin{bmatrix} 1 & 0 \\ 0 & 1 \\ 0 & 0 \end{bmatrix} = I_2$ but neither factor is invertible.

(d) False (see part (c) for an example).

(e) True (the condition forces the matrix to be square).

(f) True

21. $n \times m$

23. Let B be a right inverse. Then $AB = I$. Hence, $(AB)^T = I^T$ or $B^T A^T = I$. Hence, A^T has a left inverse, namely B^T.

25. The equivalences of (b), (c) and (d) are known. It is sufficient to show that (a) is equivalent to (b). Let us assume that A has a right inverse, say $B = [\mathbf{b}_1 \ldots \mathbf{b}_m]$. Then $AB = I$. Hence,

$$AB = [A\mathbf{b}_1 \ldots A\mathbf{b}_m] = I$$

Thus,

$$A\mathbf{b}_1 = \mathbf{e}_1, \quad \ldots, \quad A\mathbf{b}_m = \mathbf{e}_m$$

Consider now the system $A\mathbf{x} = \mathbf{b}$ for any m-vector $\mathbf{b}$. We shall show that this system is consistent. If $\mathbf{b} = (b_1, \ldots, b_m)$ then we claim that the vector $b_1\mathbf{b}_1 + \cdots + b_m\mathbf{b}_m$ is a solution to the system. Because

$$\begin{aligned} A(b_1\mathbf{b}_1 + \cdots + b_m\mathbf{b}_m) &= A(b_1\mathbf{b}_1) + \cdots + A(b_m\mathbf{b}_m) \\ &= b_1(A\mathbf{b}_1) + \cdots + b_m(A\mathbf{b}_m) \\ &= b_1\mathbf{e}_1 + \cdots + b_m\mathbf{e}_m = \mathbf{b} \end{aligned}$$

Therefore, (a) implies (b).

Conversely, if $A\mathbf{x} = \mathbf{b}$ is solvable for all $\mathbf{b}$ then for each $i = 1, \ldots, m$ the system $A\mathbf{x} = \mathbf{e}_i$ has a solution, say $\mathbf{b}_i$. Let B be the matrix with columns $\mathbf{b}_1, \ldots, \mathbf{b}_m$. Then

$$AB = [A\mathbf{b}_1 \ldots A\mathbf{b}_m] = [\mathbf{e}_1 \ldots \mathbf{e}_m] = I_m$$

Hence, A has B as its right inverse.

27. If B and C are the right and left inverse respectively, then

$$AB = I_m \text{ and } CA = I_n$$

Hence,

$$B = I_n B = (CA)B = C(AB) = CI_m = C$$

Therefore, $m = n$ and $B = C$. So, $AB = BA = I$. We conclude that A is invertible with inverse $B = C$.

Exercises 3.4

1. First we solve the system $y_1 = -11$, $-3y_1 + y_2 = 32$ to get $y_1 = -11$, $y_2 = -1$; then we solve $4x_1 + x_2 = -11$, $-x_2 = -1$ to get $x_1 = -3$, $x_2 = 1$.

3. First we solve the system $y_1 = 2$, $4y_1 + y_2 = 7$, $-2y_1 + 3y_2 + y_3 = -3$ to get $y_1 = 2$, $y_2 = -1$, $y_3 = 4$; then we solve $2x_1 - 2x_2 + x_3 = 2$, $3x_2 - x_3 = -1$, $-2x_3 = 4$ to get $x_1 = 1$, $x_2 = -1$, $x_3 = -2$.

5. First we solve the system $y_1 = 6$, $3y_1 + y_2 = 22$, $-4y_1 + 2y_2 + y_3 = -13$ to get $y_1 = 6$, $y_2 = 4$, $y_3 = 3$; then we solve $4x_1 + x_2 + x_3 = 6$, $5x_2 - x_3 = 4$, $3x_3 = 3$ to get $x_1 = 1$, $x_2 = 1$, $x_3 = 1$.

7. $LU = \begin{bmatrix} 1 & 0 \\ -5 & 1 \end{bmatrix} \begin{bmatrix} 2 & 1 \\ 0 & -7 \end{bmatrix}$

9. $LU = \begin{bmatrix} 1 & 0 & 0 \\ -4 & 1 & 0 \\ 7 & -3 & 1 \end{bmatrix} \begin{bmatrix} -1 & 2 & 1 \\ 0 & 3 & -1 \\ 0 & 0 & -5 \end{bmatrix}$

11. $LU = \begin{bmatrix} 1 & 0 & 0 & 0 \\ -3 & 1 & 0 & 0 \\ 0 & -2 & 1 & 0 \\ 5 & -1 & 0 & 1 \end{bmatrix} \begin{bmatrix} 4 & 1 & 1 & 2 \\ 0 & 2 & -1 & 2 \\ 0 & 0 & 3 & 2 \\ 0 & 0 & 0 & -1 \end{bmatrix}$

13. $LU = \begin{bmatrix} 1 & 0 & 0 & 0 \\ -3 & 1 & 0 & 0 \\ 0 & -2 & 1 & 0 \\ 5 & -1 & 0 & 1 \end{bmatrix} \begin{bmatrix} 4 & 1 & 1 \\ 0 & 2 & -1 \\ 0 & 0 & 3 \\ 0 & 0 & 0 \end{bmatrix}$

15. Using $LU = \begin{bmatrix} 1 & 0 \\ 7 & 1 \end{bmatrix} \begin{bmatrix} 2 & 1 \\ 0 & -5 \end{bmatrix}$, we get $\mathbf{x} = \begin{bmatrix} -2 \\ 10 \end{bmatrix}$.

17. Using $LU = \begin{bmatrix} 1 & 0 & 0 & 0 \\ 0 & 1 & 0 & 0 \\ -2 & 2 & 1 & 0 \\ 0 & 5 & 0 & 1 \end{bmatrix} \begin{bmatrix} 2 & 1 & 1 & 2 \\ 0 & 1 & -1 & 2 \\ 0 & 0 & 3 & 2 \\ 0 & 0 & 0 & 2 \end{bmatrix}$, we get $\mathbf{x} = \begin{bmatrix} 1 \\ 1 \\ 0 \\ 0 \end{bmatrix}$.

19. $\begin{bmatrix} 0 & 1 & 0 \\ 1 & 0 & 0 \\ 0 & 0 & 1 \end{bmatrix} A = \begin{bmatrix} 1 & 0 & 0 \\ 0 & 1 & 0 \\ -2 & -1 & 1 \end{bmatrix} \begin{bmatrix} -1 & 2 & -4 \\ 0 & 1 & 1 \\ 0 & 0 & -6 \end{bmatrix}$

21. A $PA = LU$ factorization is given by

$$\begin{bmatrix} 0 & 1 & 0 \\ 1 & 0 & 0 \\ 0 & 0 & 1 \end{bmatrix} A = \begin{bmatrix} 1 & 0 & 0 \\ 0 & 1 & 0 \\ 1 & -2 & 1 \end{bmatrix} \begin{bmatrix} 2 & 0 & 1 \\ 0 & 3 & -1 \\ 0 & 0 & -2 \end{bmatrix}$$

Hence, the LU method on the system

$$PA\mathbf{x} = P\mathbf{b} = \begin{bmatrix} -1 \\ -3 \\ -1 \end{bmatrix}$$

yields $\mathbf{x} = (-2, 0, 3)$.

23. A $PA = LU$ factorization is given by

$$\begin{bmatrix} 0 & 0 & 1 \\ 0 & 1 & 0 \\ 1 & 0 & 0 \end{bmatrix} A = \begin{bmatrix} 1 & 0 & 0 \\ 0 & 1 & 0 \\ 0 & \frac{1}{2} & 1 \end{bmatrix} \begin{bmatrix} 2 & -5 & 1 \\ 0 & 2 & -4 \\ 0 & 0 & 3 \end{bmatrix}$$

Hence, the LU method on the system

$$PA\mathbf{x} = P\mathbf{b} = \begin{bmatrix} -8 \\ 4 \\ 2 \end{bmatrix}$$

yields $\mathbf{x} = (1, 2, 0)$.

25. Let A and B be $n \times n$ unit lower triangular. Their product AB is lower triangular by Exercise 24. So, it suffices to show that the (i, i) entry of AB is 1. Because A and B are unit lower triangular

$$a_{ij} = 0, \quad b_{ij} = 0 \quad \text{for } j > i$$

$$a_{ii} = 1, \quad b_{ii} = 1$$

Let c_{ii} bet the (i, i) entry of AB. Then

$$c_{ii} = \sum_{k=1}^{n} a_{ik} b_{ki} = \sum_{k=1}^{i-1} a_{ik} b_{ki} + a_{ii} b_{ii} + \sum_{k=i+1}^{n} a_{ik} b_{ki} = 0 + 1 + 0 = 1$$

because $b_{ki} = 0$ for $k < i$ and $a_{ik} = 0$ for $k > i$ and $a_{ii} = b_{ii} = 1$. Therefore, AB is unit lower triangular.

27. Let A be an $n \times n$ invertible lower triangular matrix. Let B be its inverse. We shall prove that B is lower triangular. We know that the inverse of A^T is B^T. Because A^T is upper triangular, its inverse B^T is upper triangular (in each stage of the reduction of $[A^T : I]$ there are only zeros below the main diagonal of the left side). Hence, B is lower triangular.

29. Let A be invertible with an LU factorization, say $A = LU$. We shall prove that in this case the LU factorization is unique. In other words we shall prove that if $L'U'$ is another LU factorization then $L = L'$ and $U = U'$. First we note that L has to be invertible, or else $A = LU$ would be noninvertible. So, by Exercise 27, L^{-1} is unit lower triangular. Therefore,

$$A = LU = L'U' \Rightarrow U = (L^{-1}L')U'$$

But $L^{-1}L'$ is unit lower triangular by Exercise 25. Now the product of the lower triangular $L^{-1}L'$ with the upper triangular U' cannot be upper triangular ($= U$), unless $L^{-1}L'$ is a diagonal matrix. But because it is unit triangular it has to the identity. Hence,

$$L^{-1}L' = I \Rightarrow L = L'$$

Therefore,

$$A = LU = LU' \Rightarrow L^{-1}LU = L^{-1}LU' \Rightarrow U = U'$$

So, the LU decomposition of A is unique.

Exercises 3.5

1. The respective matrices are

$$M_1 = \begin{bmatrix} 0 & 1 \\ 1 & 0 \end{bmatrix}, \quad M_2 = \begin{bmatrix} 0 & 1 & 1 \\ 1 & 0 & 1 \\ 1 & 1 & 0 \end{bmatrix}, \quad M_3 = \begin{bmatrix} 0 & 0 & 0 \\ 0 & 0 & 0 \\ 0 & 0 & 0 \end{bmatrix}$$

3. The adjacency matrices are

$$A(G_1) = \begin{bmatrix} 0 & 1 \\ 1 & 0 \end{bmatrix}, \quad A(G_2) = \begin{bmatrix} 0 & 1 & 1 \\ 1 & 0 & 1 \\ 1 & 1 & 0 \end{bmatrix}, \quad A(G_3) = \begin{bmatrix} 0 & 0 & 0 \\ 0 & 0 & 0 \\ 0 & 0 & 0 \end{bmatrix}$$

5. We have

$$A(G_7) = \begin{bmatrix} 0 & 1 & 0 \\ 1 & 0 & 1 \\ 0 & 1 & 0 \end{bmatrix}, \quad A(G_7)^2 = \begin{bmatrix} 1 & 0 & 1 \\ 0 & 2 & 0 \\ 1 & 0 & 1 \end{bmatrix}$$

The number of walks of length 2 from (a) 1 to 1 is 1, (b) 1 to 2 is 0, (c) 1 to 3 is 1. These numbers agree with the $(1,1), (1,2), (1,3)$ entries of $A(G_7)^2$.

7. We have

$$I(G_1) = \begin{bmatrix} 1 \\ 1 \end{bmatrix}, \quad I(G_2) = \begin{bmatrix} 1 & 0 & 1 \\ 1 & 1 & 0 \\ 0 & 1 & 1 \end{bmatrix}, \quad I(G_3) = \text{Does not exist}$$

9. The graphs are

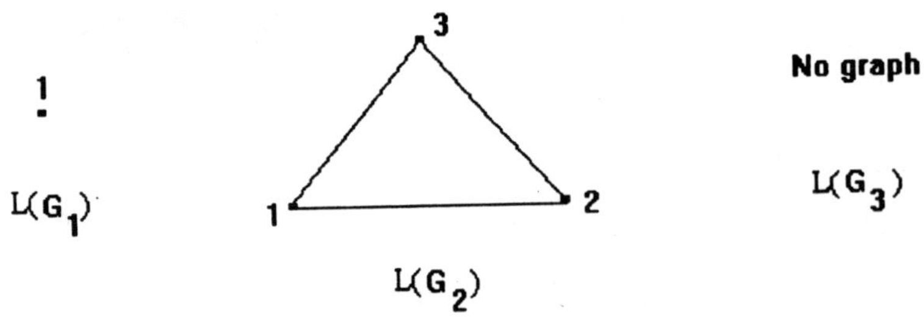

11. We have

$$A(L(G_7)) = \begin{bmatrix} 0 & 1 \\ 1 & 0 \end{bmatrix}, \qquad I(G_7) = \begin{bmatrix} 1 & 0 \\ 1 & 1 \\ 0 & 1 \end{bmatrix}$$

Hence,

$$I(G_7)^T I(G_7) - 2I_2 = \begin{bmatrix} 1 & 1 & 0 \\ 0 & 1 & 1 \end{bmatrix} \begin{bmatrix} 1 & 0 \\ 1 & 1 \\ 0 & 1 \end{bmatrix} - 2 \begin{bmatrix} 1 & 0 \\ 0 & 1 \end{bmatrix}$$

$$= \begin{bmatrix} 2 & 1 \\ 1 & 2 \end{bmatrix} - \begin{bmatrix} 2 & 0 \\ 0 & 2 \end{bmatrix} = \begin{bmatrix} 0 & 1 \\ 1 & 0 \end{bmatrix} = A(L(G_7))$$

13. Only C is stochastic. (D is not stochastic because it is not square.)

15. G and H are doubly stochastic.

17. All powers of stochastic are doubly stochastic. Therefore, the given matrix is doubly stochastic because $\begin{bmatrix} \frac{1}{2} & \frac{1}{2} \\ \frac{1}{2} & \frac{1}{2} \end{bmatrix}$ is.

19. For A we have $x = 0.8$, $y = 0.8$. For B we have $x = \frac{3}{4}$, $y = \frac{1}{4}$.

21. (a) Because

$$(I - C)^{-1} = \begin{bmatrix} 0.5 & -0.4 \\ -0.1 & 0.4 \end{bmatrix}^{-1} = \begin{bmatrix} 2.5 & 2.5 \\ 0.625 & 3.125 \end{bmatrix}$$

C is productive.

(b) The production vector is computed by

$$X = (I - C)^{-1}D = \begin{bmatrix} 2.5 & 2.5 \\ 0.625 & 3.125 \end{bmatrix} \begin{bmatrix} 10 \\ 20 \end{bmatrix} = \begin{bmatrix} 75 \\ 68.75 \end{bmatrix}$$

Exercises 4.1

1. True, because both $\begin{bmatrix} a \\ 0 \end{bmatrix} + \begin{bmatrix} b \\ 0 \end{bmatrix} = \begin{bmatrix} a+b \\ 0 \end{bmatrix}$ and $c\begin{bmatrix} a \\ 0 \end{bmatrix} = \begin{bmatrix} ca \\ 0 \end{bmatrix}$ are in V. Hence, V is a vector subspace of $\mathbf{R}^2$.

3. We have

$$\begin{bmatrix} a \\ 0 \\ -2a \end{bmatrix} + \begin{bmatrix} b \\ 0 \\ -2b \end{bmatrix} = \begin{bmatrix} a+b \\ 0 \\ -2a-2b \end{bmatrix} = \begin{bmatrix} a+b \\ 0 \\ -2(a+b) \end{bmatrix} \in V$$

and

$$c\begin{bmatrix} a \\ 0 \\ -2a \end{bmatrix} = \begin{bmatrix} ca \\ 0 \\ -2(ca) \end{bmatrix} \in V$$

Hence, V is a vector subspace of $\mathbf{R}^3$.

5. We have

$$\begin{bmatrix} a-b \\ b-c \\ c-d \\ d-a \end{bmatrix} + \begin{bmatrix} a'-b' \\ b'-c' \\ c'-d' \\ d'-a' \end{bmatrix} = \begin{bmatrix} a-b+a'-b' \\ b-c+b'-c' \\ c-d+c'-d' \\ d-a+d'-a' \end{bmatrix}$$

$$= \begin{bmatrix} (a+a')-(b+b') \\ (b+b')-(c+c') \\ (c+c')-(d+d') \\ (d+d')-(a+a') \end{bmatrix} \in V$$

Also,

$$k\begin{bmatrix} a-b \\ b-c \\ c-d \\ d-a \end{bmatrix} = \begin{bmatrix} (ka)-(kb) \\ (kb)-(kc) \\ (kc)-(kd) \\ (kd)-(ka) \end{bmatrix} \in V$$

Hence, V is a vector subspace of $\mathbf{R}^4$.

7. Let V be the set of all 4-vectors with first three components zero and let $\mathbf{v}_1$ and $\mathbf{v}_2$ be in V. Then the sum $\mathbf{v}_1 + \mathbf{v}_2$ has the first three components zero so, it is in V. Also, for any scalar c, the vector $c\mathbf{v}_1$ has the first three components zero so, it is in V. Hence, V is a vector subspace of $\mathbf{R}^4$.

9. Let $\mathbf{v}_1 = (1, -1, 0)$ and $\mathbf{v}_2 = (2, 0, 2)$ and let V be the set of all linear combinations of $\mathbf{v}_1$ and $\mathbf{v}_2$. If $\mathbf{u}_1 = c_1\mathbf{v}_1 + c_2\mathbf{v}_2$ and $\mathbf{u}_2 = k_1\mathbf{v}_1 + k_2\mathbf{v}_2$, then

$$\mathbf{u}_1 + \mathbf{u}_2 = (c_1\mathbf{v}_1 + c_2\mathbf{v}_2) + (k_1\mathbf{v}_1 + k_2\mathbf{v}_2) = (c_1 + k_1)\mathbf{v}_1 + (c_2 + k_2)\mathbf{v}_2 \in V$$

and

$$c\mathbf{u}_1 = c(c_1\mathbf{v}_1 + c_2\mathbf{v}_2) = (cc_1)\mathbf{v}_1 + (cc_2)\mathbf{v}_2 \in V$$

Hence, V is a subspace of $\mathbf{R}^3$.

11. Yes. The set is the span of $\left\{\begin{bmatrix} 1 \\ 1 \end{bmatrix}\right\}$.

13. No. The set is not closed under addition or scalar multiplication. For example, $2\begin{bmatrix} 2 \\ 0 \end{bmatrix} = \begin{bmatrix} 4 \\ 0 \end{bmatrix}$ is not in the set.

15. No. The set is not closed under addition or scalar multiplication. For example, $2\begin{bmatrix} 1 \\ 2 \end{bmatrix} = \begin{bmatrix} 2 \\ 4 \end{bmatrix}$ is not in the set.

17. Yes. This set is the span of $\left\{\begin{bmatrix} -3 \\ 1 \end{bmatrix}\right\}$.

19. No. The zero vector is not in the set; for if $\begin{bmatrix} 0 \\ 0 \end{bmatrix} = \begin{bmatrix} a-1 \\ 2a+1 \end{bmatrix}$, then $a = 1$ and $a = -\frac{1}{2}$, which is a contradiction.

21. Yes. The set is clearly closed under addition and scalar multiplication.

23. No. This set is not closed under scalar multiplication. For example, $-1\begin{bmatrix} 1 \\ 1 \end{bmatrix} = \begin{bmatrix} -1 \\ -1 \end{bmatrix}$ is not in the first quadrant.

25. No. This set is not closed under scalar multiplication. For example, $-1\begin{bmatrix}1\\1\\1\end{bmatrix} = \begin{bmatrix}-1\\-1\\-1\end{bmatrix}$ is not in the first octant.

27. Yes. This line is the span of $\left\{\begin{bmatrix}1\\1\end{bmatrix}\right\}$.

29. No. This plane does not pass through the origin.

31. $V_1 \cap V_2$ is a line or a plane (if $V_1 = V_2$) through the origin, hence, a subspace of $\mathbf{R}^3$.

33. Yes. Two linearly independent vectors span $\mathbf{R}^2$.

35. No. The set is linearly dependent.

37. Yes. Three linearly independent vectors span $\mathbf{R}^3$.

39. Yes. We have four linearly independent 4-vectors.

41. $\left\{\begin{bmatrix}1\\2\end{bmatrix}, \begin{bmatrix}-1\\1\end{bmatrix}\right\}$

43. $\left\{\begin{bmatrix}1\\0\\0\end{bmatrix}, \begin{bmatrix}0\\1\\0\end{bmatrix}, \begin{bmatrix}-1\\1\\5\end{bmatrix}\right\}$, or even the standard basis of $\mathbf{R}^3$.

45. Yes. We have two linearly independent 2-vectors.

47. No. The vectors are linearly dependent.

49. No. The vectors are linearly independent. The sum of the second and third vectors yields the fourth.

51. (a) No (b) No

53. (a) No (b) No

55. $\left\{\begin{bmatrix}1\\-1\end{bmatrix}, \begin{bmatrix}-2\\0\end{bmatrix}\right\}$

57. $\left\{ \begin{bmatrix} -5 \\ -1 \\ 1 \end{bmatrix}, \begin{bmatrix} 4 \\ 1 \\ 7 \end{bmatrix} \right\}$

59. Let $\mathbf{v}_1 \neq \mathbf{0}$ be in V. If $\text{Span}\{\mathbf{v}_1\} = V$ we take $\mathcal{B} = \{\mathbf{v}_1\}$ and stop. Otherwise, we pick a $\mathbf{v}_2 \in V$ that is not in $\text{Span}\{\mathbf{v}_1\}$. Now $\{\mathbf{v}_1, \mathbf{v}_2\}$ is linearly independent — otherwise $\mathbf{v}_2$ would a multiple of $\mathbf{v}_1$, hence it would in the span of $\mathbf{v}_1$. If $\text{Span}\{\mathbf{v}_1, \mathbf{v}_2\} = V$ we take $\mathcal{B} = \{\mathbf{v}_1, \mathbf{v}_2\}$ and stop. Otherwise, we continue in the same manner. This process terminates after we find $\{\mathbf{v}_1, \ldots \mathbf{v}_k\}$ with $k \leq n$, or else we would have more than n linearly independent vectors in $\mathbf{R}^n$, which is impossible.

61. Because $-2\begin{bmatrix} 2 \\ 1 \\ 1 \end{bmatrix} + 3\begin{bmatrix} -1 \\ 0 \\ 0 \end{bmatrix} + 4\begin{bmatrix} 0 \\ 2 \\ 1 \end{bmatrix} = \begin{bmatrix} -7 \\ 6 \\ 2 \end{bmatrix}$, we have $\mathbf{x} = \begin{bmatrix} -7 \\ 6 \\ 2 \end{bmatrix}$.

63. Because $\begin{bmatrix} 2 & -1 & 17 \\ 1 & 1 & 4 \end{bmatrix} \sim \begin{bmatrix} 1 & 0 & 7 \\ 0 & 1 & -3 \end{bmatrix}$, we have $[\mathbf{x}]_{\mathcal{B}} = \begin{bmatrix} 7 \\ -3 \end{bmatrix}$.

65. Because $\begin{bmatrix} 2 & -1 & 0 & -8 \\ 2 & 2 & -2 & 6 \\ 1 & 0 & -1 & -1 \end{bmatrix} \sim \begin{bmatrix} 1 & 0 & 0 & -2 \\ 0 & 1 & 0 & 4 \\ 0 & 0 & 1 & -1 \end{bmatrix}$, we have $[\mathbf{x}]_{\mathcal{B}} = \begin{bmatrix} -2 \\ 4 \\ -1 \end{bmatrix}$.

67. Because $\begin{bmatrix} 2 & 5 & 7a+3b \\ 1 & 0 & a-b \end{bmatrix} \sim \begin{bmatrix} 1 & 0 & a-b \\ 0 & 1 & a+b \end{bmatrix}$, we have $[\mathbf{x}]_{\mathcal{B}} = \begin{bmatrix} a-b \\ a+b \end{bmatrix}$.

Exercises 4.2

---

1. If f and g are two polynomials of degree at most n, so is the sum $f+g$ and so is the scalar product cf. Therefore, axioms (A1) and (M1) hold. Axioms (A2), (A3), (M2)–(M5) hold because the corresponding properties hold for the coefficients of the polynomials. Now P_n contains the zero polynomial by definition, and certainly $f + 0 = 0 + f = f$. Also, $-f$ has degree at most n, and $f + (-f) = 0$. Therefore, P_n is a vector space.

3. Let V be the given set and let $f = ap + bq$ and $f' = a'p + b'q$ be two elements of its elements. Then

$$f + f' = ap + bq + a'p + b'q = (a + a')p + (b + b')q \in V$$

Also,

$$cf = c(ap + bq) = (ca)p + (cb)q \in V$$

Hence, (A1) and (M1) hold. The remaining of the axioms hold, because they hold in the larger set P.

5. No, axiom (M5) fails, because $1(1,1) = (0,0) \neq (1,1)$.

7. The sum of two invertible matrices may not be invertible. For example, $\begin{bmatrix} 1 & 0 \\ 0 & 1 \end{bmatrix} + \begin{bmatrix} 0 & 1 \\ 1 & 0 \end{bmatrix} = \begin{bmatrix} 1 & 1 \\ 1 & 1 \end{bmatrix}$. Hence, axiom (A1) is not satisfied.

9. No; for example, $2p$ has leading coefficient 2 not 1 as required. So, the set is not closed under scalar multiplication.

11. No. If f and g have coefficients that add up to 1, then the coefficients of $f + g$ add up to 2. So, $f + g$ does *not* belong to the set.

13. Yes. This is the same as Exercise 8.

15. No. The sum $(1 + x^2) + (x + x^2) = 2x^2 + x + 1$ is not in the set.

17. Yes, because if V is the set, then

$$\begin{bmatrix} a & -b \\ b & a \end{bmatrix} + \begin{bmatrix} a' & -b' \\ b' & a' \end{bmatrix} = \begin{bmatrix} a + a' & -(b + b') \\ b + b' & a + a' \end{bmatrix} \in V$$

and

$$c\begin{bmatrix} a & -b \\ b & a \end{bmatrix} = \begin{bmatrix} ca & -(cb) \\ cb & ca \end{bmatrix} \in V$$

19. No, the zero matrix is not in the set.

21. Yes. The verification of all axioms is straightforward.

23. Yes. All axioms can be easily verified.

25. Yes. If f and g are even, so is the sum $f+g$, because

$$(f+g)(-x) = f(-x) + g(-x) = f(x) + g(x) = (f+g)(x)$$

Also cf is even, because

$$(cf)(-x) = cf(-x) = cf(x) = (cf)(x)$$

27. 2. $c\mathbf{0} = c(\mathbf{u} + (-\mathbf{u})) = c\mathbf{u} + c(-\mathbf{u}) = c\mathbf{u} + (-(c\mathbf{u})) = \mathbf{0}$

3. If $c = 0$, we are done. If $c \neq 0$, then

$$\begin{aligned} c\mathbf{u} = \mathbf{0} \quad &\Rightarrow \quad c^{-1}(c\mathbf{u}) = c^{-1}\mathbf{0} \\ &\Rightarrow \quad (c^{-1}c)\mathbf{u} = \mathbf{0} \\ &\Rightarrow \quad 1\mathbf{u} = \mathbf{0} \\ &\Rightarrow \quad \mathbf{u} = \mathbf{0} \end{aligned}$$

29. If $r = 0$, we are done. If $r \neq 0$, then r^{-1} exists and

$$\begin{aligned} r\mathbf{u} = r\mathbf{v} \quad &\Rightarrow \quad r^{-1}(r\mathbf{u}) = r^{-1}(r\mathbf{v}) \\ &\Rightarrow \quad (r^{-1}r)\mathbf{u} = (r^{-1}r)\mathbf{v} \\ &\Rightarrow \quad 1\mathbf{u} = 1\mathbf{v} \\ &\Rightarrow \quad \mathbf{u} = \mathbf{v} \end{aligned}$$

31. Let $\mathbf{v}_1$ and $\mathbf{v}_2$ be n–vectors in $U \cap W$. It suffices to show that $\mathbf{v}_1 + \mathbf{v}_2$ and $c\mathbf{v}_1$ are in $U \cap W$. Vectors $\mathbf{v}_1 + \mathbf{v}_2$ and $c\mathbf{v}_1$ are in U because $\mathbf{v}_1$ and $\mathbf{v}_2$ are in U, and U is a subspace. Similarly, $\mathbf{v}_1 + \mathbf{v}_2$ and $c\mathbf{v}_1$ are in W. Hence, $\mathbf{v}_1 + \mathbf{v}_2$ and $c\mathbf{v}_1$ are in the intersection $U \cap W$.

Exercises 4.3

1. (a) Yes. $4-x^2=c_1(1+3x-x^2)+c_2(-3x+2x^2)+c_3x^2$ implies

$$4-x^2=c_1+(3c_1-3c_2)\,x+(-c_1+2c_2+c_3)\,x^2$$

So, $c_1=4,\ 3c_1-3c_2=0,\ -c_1+2c_2+c_3=-1$, or $c_1=4,\ c_2=4,\ c_3=-5$. Therefore, p_4 is a linear combination of the other three polynomials.

(b) No. Because $4-x^2=c_1(1+3x-x^2)+c_2(-3x+2x^2)$ implies

$$4-x^2=c_1+(3c_1-3c_2)\,x+(-c_1+2c_2)\,x^2$$

Hence, $c_1=4,\ 3c_1-3c_2=0,\ -c_1+2c_2=-1$. But this system is inconsistent, so p_4 is not a linear combination of $\{p_1,p_2\}$.

(c) No. The equation $4-x^2=c_1(-3x+2x^2)+c_2x^2$ is impossible, because the constant terms on the two sides are different. So, p_4 is not a linear combination of $\{p_2,p_3\}$.

3. No. For any polynomial $a+bx+cx^2$ we want to find scalars c_1,c_2,c_3 such that

$$c_1\left(1+x+x^2\right)+c_2\left(1+2x+x^2\right)+c_3x=a+bx+cx^2$$

Thus,

$$(c_1+c_2)+(c_1+2c_2+c_3)\,x+(c_1+c_2)\,x^2=a+bx+cx^2$$

We need to show that the following system is consistent for all choices of a,b,c

$$\begin{aligned} c_1+c_2 &= a\\ c_1+2c_2+c_3 &= b\\ c_1+c_2 &= c \end{aligned}$$

This is not true, because the coefficient matrix has only two pivots as we see from

$$\begin{bmatrix}1&1&0\\1&2&1\\1&1&0\end{bmatrix}\sim\begin{bmatrix}1&1&0\\0&1&1\\0&0&0\end{bmatrix}$$

5. Yes. For any polynomial $a + bx + cx^2$ we want to find scalars c_1, c_2, c_3 such that

$$c_1(1+x) + c_2(-1+x) + c_3\left(2+x+x^2\right) = a + bx + cx^2$$

Thus,

$$(c_1 - c_2 + 2c_3) + (c_1 + c_2 + c_3)x + c_3x^2 = a + bx + cx^2$$

We need to show that the system is consistent for all choices of a, b, c

$$\begin{aligned} c_1 - c_2 + 2c_3 &= a \\ c_1 + c_2 + c_3 &= b \\ c_3 &= c \end{aligned}$$

This is true, because the coefficient matrix has three pivots as we see from

$$\begin{bmatrix} 1 & -1 & 2 \\ 1 & 1 & 1 \\ 0 & 0 & 1 \end{bmatrix} \sim \begin{bmatrix} 1 & -1 & 2 \\ 0 & 2 & -1 \\ 0 & 0 & 1 \end{bmatrix}$$

7. Yes. For any polynomial $a + bx + cx^2$ we want to find scalars c_1, c_2, c_3 such that

$$c_1\left(-1+x+x^2\right) + c_2\left(1-x+x^2\right) + c_3\left(1+x-x^2\right) = a + bx + cx^2$$

or

$$(-c_1 + c_2 + c_3) + (c_1 - c_2 + c_3)x + (c_1 + c_2 - c_3)x^2 = a + bx + cx^2$$

Thus, we need to show that the system is consistent for all choices of a, b, c

$$\begin{aligned} -c_1 + c_2 + c_3 &= a \\ c_1 - c_2 + c_3 &= b \\ c_1 + c_2 - c_3 &= c \end{aligned}$$

This is true, because the coefficient matrix has three pivots as we see from

$$\begin{bmatrix} -1 & 1 & 1 \\ 1 & -1 & 1 \\ 1 & 1 & -1 \end{bmatrix} \sim \begin{bmatrix} -1 & 1 & 1 \\ 0 & 2 & 0 \\ 0 & 0 & 2 \end{bmatrix}$$

9. No. If these matrices spanned M_{22} then there should be scalars c_1, c_2, c_3, c_4 such that

$$c_1\begin{bmatrix} 1 & 1 \\ 1 & 1 \end{bmatrix} + c_2\begin{bmatrix} 0 & 3 \\ 0 & 3 \end{bmatrix} + c_3\begin{bmatrix} 4 & 0 \\ 4 & 0 \end{bmatrix} + c_4\begin{bmatrix} 0 & 5 \\ 0 & 5 \end{bmatrix} = \begin{bmatrix} a & b \\ c & b \end{bmatrix}$$

or

$$\begin{bmatrix} c_1 + 4c_3 & c_1 + 3c_2 + 5c_4 \\ c_1 + 4c_3 & c_1 + 3c_2 + 5c_4 \end{bmatrix} = \begin{bmatrix} a & b \\ c & b \end{bmatrix}$$

for all choices of a, b, c, d. However, the resulting system

$$\begin{aligned} c_1 + 4c_3 &= a \\ c_1 + 3c_2 + 5c_4 &= b \\ c_1 + 4c_3 &= c \\ c_1 + 3c_2 + 5c_4 &= d \end{aligned}$$

has coefficient matrix with linearly dependent columns (column 2 is a multiple of column 4), so it cannot have four pivots; thus, it cannot be solved for all a, b, c, d. So, the matrices do not span M_{22}.

11. Yes. In the notation of Exercise 9, we see from the reduction

$$\begin{bmatrix} 1 & 1 & 2 & 3 \\ 0 & 1 & 1 & 0 \\ 0 & 1 & 0 & -1 \\ 0 & 0 & 0 & 1 \end{bmatrix} \sim \begin{bmatrix} 1 & 1 & 2 & 3 \\ 0 & 1 & 1 & 0 \\ 0 & 0 & -1 & -1 \\ 0 & 0 & 0 & 1 \end{bmatrix}$$

that the coefficient matrix of the resulting system has four pivots, so this time the system can be solved for all choices of a, b, c, d; hence, the matrices span M_{22}.

13. Yes. The relation

$$c_1 \begin{bmatrix} 1 & 0 \\ 0 & 0 \end{bmatrix} + c_2 \begin{bmatrix} 1 & 1 \\ 0 & 0 \end{bmatrix} + c_3 \begin{bmatrix} 1 & 1 \\ 1 & 0 \end{bmatrix} + c_4 \begin{bmatrix} 1 & 1 \\ 1 & 1 \end{bmatrix} = \begin{bmatrix} a & b \\ c & d \end{bmatrix}$$

yields the system with the four-pivoted coefficient matrix

$$\begin{bmatrix} 1 & 1 & 1 & 1 \\ 0 & 1 & 1 & 1 \\ 0 & 0 & 1 & 1 \\ 0 & 0 & 0 & 1 \end{bmatrix}$$

and vector of constants (a, b, c, d). So this system can be solved for choices of $a, b, c,$ and d. Thus, these matrices span M_{22}.

15. (a) False (b) True (c) False (d) True (e) True (f) False (g) True

17. Let V_1 and V_2 be the two sets. Because $\mathbf{u}+\mathbf{v}$, $\mathbf{u}+\mathbf{v}+\mathbf{w}$ are in V_1, we see that V_2 is a subset of V_1. On the other hand, because

$$\mathbf{v}=(\mathbf{u}+\mathbf{v})-\mathbf{u}, \qquad \mathbf{w}=(\mathbf{u}+\mathbf{v}+\mathbf{w})-(\mathbf{u}+\mathbf{v})$$

we see that V_1 is a subset of V_2. Hence, $V_1=V_2$.

19. No. We have

$$c_1\left(1+ax+ax^2\right)+c_2\left(1+bx+bx^2\right)+c_3(1)=0$$

Thus,

$$(c_1+c_3+c_2)+(c_1a+c_2b)\,x+(c_1a+c_2b)\,x^2=0$$

Hence, $c_1+c_2+c_3=0$, $c_1a+c_2b=0$, $c_1a+c_2b=0$. But this system has nontrivial solutions, for any choices of a and b.

21. Yes. The relation

$$c_1\begin{bmatrix}1&0\\0&0\end{bmatrix}+c_2\begin{bmatrix}1&1\\1&0\end{bmatrix}+c_3\begin{bmatrix}2&1\\0&0\end{bmatrix}+c_4\begin{bmatrix}3&0\\-1&1\end{bmatrix}=\begin{bmatrix}0&0\\0&0\end{bmatrix}$$

implies the system $c_1+c_2+2c_3+3c_4=0$, $c_2+c_3=0$, $c_2-c_4=0$, $c_4=0$, which has only the trivial solution.

23. (a) False (b) False (c) True (d) False

25. Let

$$c_1(\mathbf{v}_1-\mathbf{v}_2)+c_2(\mathbf{v}_2-\mathbf{v}_3)+c_3(\mathbf{v}_3+\mathbf{v}_1)=\mathbf{0}$$

Then

$$(c_1+c_3)\mathbf{v}_1+(c_2-c_1)\mathbf{v}_2+(c_3-c_2)\mathbf{v}_3=\mathbf{0}$$

Hence, $c_1+c_3=0$, $c_2-c_1=0$, $c_3-c_2=0$, because $\{\mathbf{v}_1,\mathbf{v}_2,\mathbf{v}_3\}$ is linearly independent. But the last system implies that $c_1=c_2=c_3=0$. Hence, $\{\mathbf{v}_1-\mathbf{v}_2,\mathbf{v}_2-\mathbf{v}_3,\mathbf{v}_3+\mathbf{v}_1\}$ is linearly independent.

27. No. For example, take $p=x+1$, $q=1$, and $r=-x-1$.

29. No. The matrices are linearly dependent as can be seen from the reduction of the coefficient matrix of the corresponding system

$$\begin{bmatrix}1&0&1&0\\0&1&0&1\\0&1&1&0\\1&0&0&1\end{bmatrix}\sim\begin{bmatrix}1&0&1&0\\0&1&0&1\\0&0&1&-1\\0&0&0&0\end{bmatrix}$$

31. In Exercise 18, Section 4.2 we saw that V is a vector subspace of M_{22}. If we take $a = 1$, $b = c = 0$ we see that $E_{11} - E_{22}$ is in V. Also from $a = c = 0$ and $b = 1$ we see that E_{12} is in V. Finally, taking $a = b = 0$, $c = 1$ shows that E_{21} is in V. B spans V because

$$\begin{bmatrix} a & b \\ c & -a \end{bmatrix} = a\begin{bmatrix} 1 & 0 \\ 0 & -1 \end{bmatrix} + b\begin{bmatrix} 0 & 1 \\ 0 & 0 \end{bmatrix} + c\begin{bmatrix} 0 & 0 \\ 1 & 0 \end{bmatrix}$$

It is also linearly independent, because the right-hand side linear combination, when set equal to zero, implies that the left-hand side matrix is zero. Hence, $a = b = c = 0$. So, $\mathcal{B}$ is a basis of V.

33. True because,

$$\begin{bmatrix} 1 & 0 & 0 & 3 \\ 0 & 1 & 0 & -3 \\ 0 & 0 & 2 & 0 \\ 0 & 0 & 0 & 1 \end{bmatrix}$$

has exactly four pivots.

35. (a) $\mathcal{B} = \{2 + x + 2x^2, x^2, 1 - x - x^2\}$ (b) $V = P_2$

37. (a) $\mathcal{B} = \{x^2, 1 + x, -1 + x^2\}$ (b) $V = P_2$

39. (a) $\mathcal{B} = \{-x + x^2, -5 + x, -x^2\}$ (b) $V = P_2$

41. $\{-x + x^2, x + x^2, 1\}$

43. $\{1 + x, -1 + x^2, x^2\}$

45. $\{-x + x^2, -5 + x, x^2\}$

Exercises 4.4

1. $\dim(V) = 1$, because $\left\{ \begin{bmatrix} 1 \\ 0 \end{bmatrix} \right\}$ is a basis of V.

3. $\dim(V) = 1$, because $\{(1, 0, -2)\}$ is a basis of V.

5. $\dim(V) = 2$, because $\{(0, 1, 0), (0, 0, 1)\}$ is a basis of V.

7. $\dim(V) = 1$, because $\{(0, 0, 0, 1)\}$ is a basis of V.

9. The dimension is 2, because the first two matrices are a basis for the span, as seen from

$$\begin{bmatrix} 1 & 3 & 5 \\ 2 & 4 & 6 \\ 1 & 3 & 5 \\ 2 & 4 & 6 \end{bmatrix} \sim \begin{bmatrix} 1 & 3 & 5 \\ 0 & -2 & -4 \\ 0 & 0 & 0 \\ 0 & 0 & 0 \end{bmatrix}$$

11. The dimension is 3, because all matrices are linearly independent, as seen by

$$\begin{bmatrix} 1 & 2 & 3 \\ 2 & 2 & 3 \\ 3 & 3 & 3 \\ 4 & 4 & 4 \end{bmatrix} \sim \begin{bmatrix} 1 & 2 & 3 \\ 0 & -2 & -3 \\ 0 & 0 & -2 \\ 0 & 0 & 0 \end{bmatrix}$$

13. $\dim(V) = 3$, because $\{1 + x + x^2, 1, -1 - x^2\}$ is a basis, as seen by

$$\begin{bmatrix} 1 & 1 & 1 & 0 \\ 1 & 0 & 0 & 0 \\ 1 & 0 & -1 & 1 \end{bmatrix} \sim \begin{bmatrix} 1 & 1 & 1 & 0 \\ 0 & -1 & -1 & 0 \\ 0 & 0 & -1 & 1 \end{bmatrix}$$

15. $\dim(V) = 2$, because $\{x + x^2, 1 + x\}$ is a basis, as we see from the reduction

$$\begin{bmatrix} 0 & 1 & -1 \\ 1 & 1 & 0 \\ 1 & 0 & 1 \end{bmatrix} \sim \begin{bmatrix} 1 & 1 & 0 \\ 0 & 1 & -1 \\ 0 & 0 & 0 \end{bmatrix}$$

17. $\dim(V) = 2$, because $\{1 - x - 5x^2, 7 + x + 4x^2\}$ by observing that the last polynomial in the given set is the sum of the first two.

19. (a) False (b) True (c) False (d) True (e) False (f) True (g) False

21. The dimension is 2, because $\{e^x, e^{2x}\}$ is a basis of the span.

23. The dimension is 2. $\{\cos^2 x, \sin^2 x\}$ is a basis of the span because $\cos^2 x + \sin^2 x = 1$.

25. $\{0\}$, $\mathbf{R}^4$, all lines through the origin, all planes through the origin, and all hyperplanes through the origin.

Exercises 4.5

---

1. $p = -3(1 + 2x) + 6(5x) = -3 + 24x$

3. $p = a(2 + 2x) + b(-3 + 3x) = (2a - 3b) + (2a + 3b)x$

5. We have $c_1(-7 + 4x) + c_2(2 - 3x) = 17 - 6x$. Hence,

$$\begin{bmatrix} -7 & 2 & 17 \\ 4 & -3 & -6 \end{bmatrix} \sim \begin{bmatrix} 1 & 0 & -3 \\ 0 & 1 & -2 \end{bmatrix}$$

Thus, $[p]_{\mathcal{B}} = \begin{bmatrix} -3 \\ -2 \end{bmatrix}$.

7. We have $c_1(1 + 2x) + c_2(5x) = (a - b) + (7a + 3b)x$. Hence,

$$\begin{bmatrix} 1 & 0 & a - b \\ 2 & 5 & 7a + 3b \end{bmatrix} \sim \begin{bmatrix} 1 & 0 & a - b \\ 0 & 1 & a + b \end{bmatrix}$$

Therefore, $[p]_{\mathcal{B}} = \begin{bmatrix} a - b \\ a + b \end{bmatrix}$.

9. We have

$$c_1\begin{bmatrix}1 & 1\\ 0 & 0\end{bmatrix}+c_2\begin{bmatrix}-1 & 0\\ 0 & 0\end{bmatrix}+c_3\begin{bmatrix}2 & 0\\ -1 & 0\end{bmatrix}+c_4\begin{bmatrix}1 & 2\\ 3 & 4\end{bmatrix}=\begin{bmatrix}4 & -1\\ -4 & -4\end{bmatrix}$$

Hence,

$$\begin{bmatrix}1 & -1 & 2 & 1 & 4\\ 1 & 0 & 0 & 2 & -1\\ 0 & 0 & -1 & 3 & -4\\ 0 & 0 & 0 & 4 & -4\end{bmatrix}\sim\begin{bmatrix}1 & 0 & 0 & 0 & 1\\ 0 & 1 & 0 & 0 & -2\\ 0 & 0 & 1 & 0 & 1\\ 0 & 0 & 0 & 1 & -1\end{bmatrix}$$

So, the coordinate vector is $(1,-2,1,-1)$.

11. The matrix is $\begin{bmatrix}0 & 1\\ 1 & 0\end{bmatrix}$, as we see from

$$\begin{bmatrix}0 & 1 & : & 1 & 0\\ 1 & 0 & : & 0 & 1\end{bmatrix}\sim\begin{bmatrix}1 & 0 & : & 0 & 1\\ 0 & 1 & : & 1 & 0\end{bmatrix}$$

13. The matrix is $\begin{bmatrix}0 & 0 & 1\\ 1 & 0 & 0\\ 0 & 1 & 0\end{bmatrix}$, as we seen from

$$\begin{bmatrix}0 & 1 & 0 & : & 1 & 0 & 0\\ 0 & 0 & 1 & : & 0 & 1 & 0\\ 1 & 0 & 0 & : & 0 & 0 & 1\end{bmatrix}\sim\begin{bmatrix}1 & 0 & 0 & : & 0 & 0 & 1\\ 0 & 1 & 0 & : & 1 & 0 & 0\\ 0 & 0 & 1 & : & 0 & 1 & 0\end{bmatrix}$$

15. The matrix is

$$\begin{bmatrix}1 & 0 & \frac{1}{4} & 0\\ 0 & \frac{1}{2} & 0 & \frac{1}{4}\\ 0 & 0 & \frac{1}{4} & 0\\ 0 & 0 & 0 & \frac{1}{8}\end{bmatrix}$$

as we see from

$$\begin{bmatrix}1 & 0 & -1 & 0 & 1 & 0 & 0 & 0\\ 0 & 2 & 0 & -4 & 0 & 1 & 0 & 0\\ 0 & 0 & 4 & 0 & 0 & 0 & 1 & 0\\ 0 & 0 & 0 & 8 & 0 & 0 & 0 & 1\end{bmatrix}\sim\begin{bmatrix}1 & 0 & 0 & 0 & 1 & 0 & \frac{1}{4} & 0\\ 0 & 1 & 0 & 0 & 0 & \frac{1}{2} & 0 & \frac{1}{4}\\ 0 & 0 & 1 & 0 & 0 & 0 & \frac{1}{4} & 0\\ 0 & 0 & 0 & 1 & 0 & 0 & 0 & \frac{1}{8}\end{bmatrix}$$

17. The matrix is

$$\begin{bmatrix} 1 & 0 & \frac{1}{2} & 0 \\ 0 & \frac{1}{2} & 0 & \frac{3}{4} \\ 0 & 0 & \frac{1}{4} & 0 \\ 0 & 0 & 0 & \frac{1}{8} \end{bmatrix}$$

as seen by

$$\begin{bmatrix} 1 & 0 & -2 & 0 & 1 & 0 & 0 & 0 \\ 0 & 2 & 0 & -12 & 0 & 1 & 0 & 0 \\ 0 & 0 & 4 & 0 & 0 & 0 & 1 & 0 \\ 0 & 0 & 0 & 8 & 0 & 0 & 0 & 1 \end{bmatrix} \sim \begin{bmatrix} 1 & 0 & 0 & 0 & 1 & 0 & \frac{1}{2} & 0 \\ 0 & 1 & 0 & 0 & 0 & \frac{1}{2} & 0 & \frac{3}{4} \\ 0 & 0. & 1 & 0 & 0 & 0 & \frac{1}{4} & 0 \\ 0 & 0 & 0 & 1 & 0 & 0 & 0 & \frac{1}{8} \end{bmatrix}$$

19. The matrix is

$$\begin{bmatrix} 1 & \frac{1}{2} & \frac{1}{2} & \frac{1}{2} \\ 0 & 1 & 1 & \frac{3}{2} \\ 0 & 0 & 1 & \frac{3}{2} \\ 0 & 0 & 0 & 1 \end{bmatrix}$$

as we see from the reduction

$$\begin{bmatrix} 1 & -\frac{1}{2} & 0 & \frac{1}{4} & 1 & 0 & 0 & 0 \\ 0 & 1 & -1 & 0 & 0 & 1 & 0 & 0 \\ 0 & 0 & 1 & -\frac{3}{2} & 0 & 0 & 1 & 0 \\ 0 & 0 & 0 & 1 & 0 & 0 & 0 & 1 \end{bmatrix} \sim \begin{bmatrix} 1 & 0 & 0 & 0 & 1 & \frac{1}{2} & \frac{1}{2} & \frac{1}{2} \\ 0 & 1 & 0 & 0 & 0 & 1 & 1 & \frac{3}{2} \\ 0 & 0 & 1 & 0 & 0 & 0 & 1 & \frac{3}{2} \\ 0 & 0 & 0 & 1 & 0 & 0 & 0 & 1 \end{bmatrix}$$

21. Because the new basis is $\{(0,-1),(-1,0)\}$, the transition matrix is $\begin{bmatrix} 0 & -1 \\ -1 & 0 \end{bmatrix}$, as we see from

$$\begin{bmatrix} 0 & -1 & 1 & 0 \\ -1 & 0 & 0 & 1 \end{bmatrix} \sim \begin{bmatrix} 1 & 0 & 0 & -1 \\ 0 & 1 & -1 & 0 \end{bmatrix}$$

Hence, the new coordinates of $(1,1)$ are given by $\begin{bmatrix} 0 & -1 \\ -1 & 0 \end{bmatrix}\begin{bmatrix} 1 \\ 1 \end{bmatrix} = \begin{bmatrix} -1 \\ -1 \end{bmatrix}$.

Exercises 4.6

1. The reductions

$$\begin{bmatrix} -1 & 2 & 0 \\ 2 & -4 & 0 \end{bmatrix} \sim \begin{bmatrix} 1 & -2 & 0 \\ 0 & 0 & 0 \end{bmatrix}, \quad \begin{bmatrix} 2 & -2 & 2 & 0 \\ 3 & -3 & 3 & 0 \\ 4 & -4 & 5 & 0 \end{bmatrix} \sim \begin{bmatrix} 1 & -1 & 0 & 0 \\ 0 & 0 & 1 & 0 \\ 0 & 0 & 0 & 0 \end{bmatrix}$$

yield the following general solutions for each $A\mathbf{x} = \mathbf{0}$.

(a) $x_1 = 2r$, $x_2 = r$. Therefore, $\left\{ \begin{bmatrix} 2 \\ 1 \end{bmatrix} \right\}$ is a basis for the null space. The nullity is 1.

(b) $x_1 = r$, $x_2 = r, x_3 = 0$. Therefore, $\left\{ \begin{bmatrix} 1 \\ 1 \\ 0 \end{bmatrix} \right\}$ is a basis for the null space. The nullity is 1.

3. The reductions

$$\begin{bmatrix} 1 & 2 & -1 & -3 & 0 & 6 & 0 \\ 0 & 0 & 0 & 0 & 2 & 4 & 0 \\ 0 & 0 & 0 & 0 & 0 & 9 & 0 \end{bmatrix} \sim \begin{bmatrix} 1 & 2 & -1 & -3 & 0 & 0 & 0 \\ 0 & 0 & 0 & 0 & 1 & 0 & 0 \\ 0 & 0 & 0 & 0 & 0 & 1 & 0 \end{bmatrix}$$

$$\begin{bmatrix} -1 & 1 & 1 & 0 \\ 0 & 2 & -2 & 0 \\ 0 & 0 & 3 & 0 \\ 0 & 0 & 1 & 0 \end{bmatrix} \sim \begin{bmatrix} 1 & 0 & 0 & 0 \\ 0 & 1 & 0 & 0 \\ 0 & 0 & 1 & 0 \\ 0 & 0 & 0 & 0 \end{bmatrix}$$

yield the following for each matrix.

(a) A null space basis is $\{(-2,1,0,0,0,0), (1,0,1,0,0,0), (3,0,0,1,0,0)\}$. Nullity: 3.

(b) The only null space basis is the empty set. The nullity is 0.

5. The reductions

$$\begin{bmatrix} 1 & -1 & 2 & -1 & 0 \\ -1 & 0 & -1 & 2 & 0 \\ 2 & -4 & 6 & 0 & 0 \end{bmatrix} \sim \begin{bmatrix} 1 & 0 & 1 & -2 & 0 \\ 0 & 1 & -1 & -1 & 0 \\ 0 & 0 & 0 & 0 & 0 \end{bmatrix}$$

$$\begin{bmatrix} 1 & -1 & 0 \\ -1 & 1 & 0 \\ 1 & -1 & 0 \\ -4 & 4 & 0 \\ 0 & 0 & 0 \end{bmatrix} \sim \begin{bmatrix} 1 & -1 & 0 \\ 0 & 0 & 0 \\ 0 & 0 & 0 \\ 0 & 0 & 0 \\ 0 & 0 & 0 \end{bmatrix}$$

yield the following.

(a) A null space basis is $\{(-1,1,1,0), (2,1,0,1)\}$. The nullity is 2.

(b) A null space basis is $\left\{\begin{bmatrix} 1 \\ 1 \end{bmatrix}\right\}$. The nullity is 1.

7. The reduction

$$\begin{bmatrix} 1 & -1 & 2 & 3 & 0 & 0 \\ 2 & -1 & 6 & 0 & 1 & 0 \\ -1 & 2 & 0 & -1 & 1 & 0 \end{bmatrix} \sim \begin{bmatrix} 1 & 0 & 4 & 0 & 1 & 0 \\ 0 & 1 & 2 & 0 & 1 & 0 \\ 0 & 0 & 0 & 1 & 0 & 0 \end{bmatrix}$$

shows that a null space basis is $\{(-4,-2,1,0,0), (-1,-1,0,0,1)\}$. Therefore, the nullity is 2.

Let N be the nullity, let P be the number of pivot columns, and let C be the number of columns of a matrix.

9. As we see from the reductions

$$\begin{bmatrix} 1 & 0 & 1 & -1 \\ 0 & 2 & 2 & -2 \end{bmatrix} \sim \begin{bmatrix} 1 & 0 & 1 & -1 \\ 0 & 1 & 1 & -1 \end{bmatrix}, \quad \begin{bmatrix} 1 & 1 & 2 \\ 0 & 0 & 0 \\ 0 & 1 & 1 \\ 0 & -1 & -1 \end{bmatrix} \sim \begin{bmatrix} 1 & 0 & 1 \\ 0 & 1 & 1 \\ 0 & 0 & 0 \\ 0 & 0 & 0 \end{bmatrix}$$

(a) $N = 2$, $P = 2$, $2 + 2 = 4 = C$ (b) $N = 1$, $P = 2$, $1 + 2 = 3 = C$

11. Because the solution space is a subspace, for any two solutions $\mathbf{x}_1$, $\mathbf{x}_2$, the sum $\mathbf{x}_1 + \mathbf{x}_2$ must be a solution. So, $\mathbf{b} = A(\mathbf{x}_1 + \mathbf{x}_2) = A\mathbf{x}_1 + A\mathbf{x}_2 = \mathbf{b} + \mathbf{b} = 2\mathbf{b}$, because $A\mathbf{x}_1 = \mathbf{b}$ and $A\mathbf{x}_2 = \mathbf{b}$. Therefore, $\mathbf{b} = \mathbf{0}$.

13. Vectors **a**, **b**, **c** are in the column space, as we can see by the common reduction

$$\begin{bmatrix} 1 & 2 & 3 & -2 & 3 & -5 \\ 4 & 5 & 6 & 4 & -6 & 7 \end{bmatrix} \sim \begin{bmatrix} 1 & 2 & 3 & -2 & 3 & -5 \\ 0 & -3 & -6 & 12 & -18 & 27 \end{bmatrix}$$

Vectors **u**, **v**, **w** are the wrong size.

15. Only **u** and **w** are in the column space, because only the **v**-column is pivot in

$$\begin{bmatrix} 1 & -2 & 3 & 1 & -4 & 3 \\ 0 & -4 & 5 & 2 & -8 & -1 \\ 0 & 0 & 0 & 0 & 1 & 0 \end{bmatrix}$$

and **a**, **b**, **c** are the wrong size.

17. None of the vectors is in the column space. Firstly, vectors **a**, **b**, **c** are the wrong size. Secondly, vectors **u**, **v**, **w** correspond to pivot columns in

$$\begin{bmatrix} -2 & 1 & 4 & -6 & 1 & 1 & -4 & 3 \\ 0 & 4 & -5 & 0 & 2 & 2 & -8 & -1 \\ 0 & -8 & 10 & 0 & -4 & 0 & 1 & 0 \end{bmatrix} \sim \begin{bmatrix} -2 & 1 & 4 & -6 & 1 & 1 & -4 & 3 \\ 0 & 4 & -5 & 0 & 2 & 2 & -8 & -1 \\ 0 & 0 & 0 & 0 & 0 & 4 & -15 & -2 \end{bmatrix}$$

19. $\left\{ \begin{bmatrix} 2 \\ 0 \\ 0 \end{bmatrix}, \begin{bmatrix} 0 \\ 1 \\ 0 \end{bmatrix}, \begin{bmatrix} 0 \\ 1 \\ 1 \end{bmatrix} \right\}$

21. $\left\{ \begin{bmatrix} 1 \\ 0 \\ 0 \\ 0 \end{bmatrix}, \begin{bmatrix} -2 \\ -2 \\ 2 \\ 0 \end{bmatrix}, \begin{bmatrix} 1 \\ 2 \\ -4 \\ 0 \end{bmatrix}, \begin{bmatrix} 0 \\ 4 \\ 2 \\ 2 \end{bmatrix} \right\}$, as we see from the reduction

$$\begin{bmatrix} 1 & -2 & 1 & 0 & 0 & 0 & 0 \\ 0 & -2 & 2 & 2 & 4 & 6 & -5 \\ 0 & 2 & -4 & 0 & 2 & 1 & 2 \\ 0 & 0 & 0 & 0 & 2 & 1 & 2 \end{bmatrix} \sim \begin{bmatrix} 1 & -2 & 1 & 0 & 0 & 0 & 0 \\ 0 & -2 & 2 & 2 & 4 & 6 & -5 \\ 0 & 0 & -2 & 2 & 6 & 7 & -3 \\ 0 & 0 & 0 & 0 & 2 & 1 & 2 \end{bmatrix}$$

23. $\left\{ \begin{bmatrix} 1 \\ 0 \\ 0 \\ 0 \end{bmatrix}, \begin{bmatrix} -2 \\ -2 \\ 0 \\ 0 \end{bmatrix}, \begin{bmatrix} 0 \\ 4 \\ 2 \\ 2 \end{bmatrix} \right\}$, as seen by a one-step reduction.

25. The column space is $\mathbf{R}^2$.

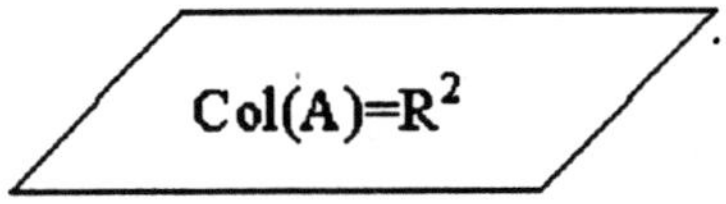

27. The null space is spanned by $\{(0,2,1)\}$ and the column space is spanned by the set $\{(3,0,0),(0,1,-2)\}$.

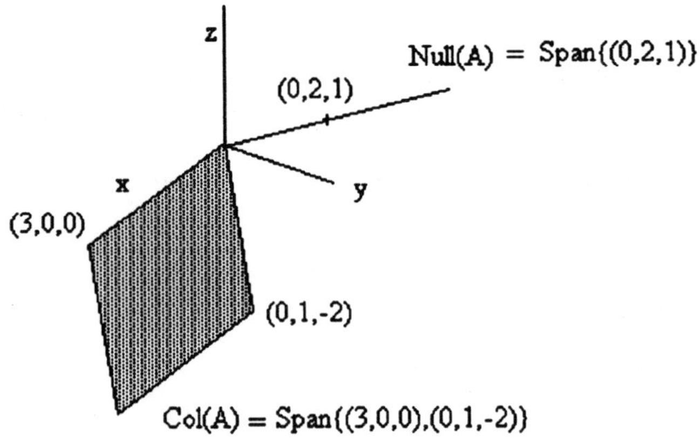

29. $A = \begin{bmatrix} 1 & 1 \\ 1 & 1 \end{bmatrix}$ and $B = \begin{bmatrix} 1 & 1 \\ 0 & 0 \end{bmatrix}$. We have $A \sim B$. However, Col(A) consists of all multiples of $\begin{bmatrix} 1 \\ 1 \end{bmatrix}$, whereas Col$(B)$ consists of all multiples of $\begin{bmatrix} 1 \\ 0 \end{bmatrix}$.

31. $\left\{ \begin{bmatrix} -1 \\ 2 \\ 3 \end{bmatrix}, \begin{bmatrix} 0 \\ 1 \\ 1 \end{bmatrix} \right\}$, because $\begin{bmatrix} -1 & 3 & 0 & 3 \\ 2 & -6 & 1 & -5 \\ 3 & -9 & 1 & -8 \end{bmatrix} \sim \begin{bmatrix} -1 & 3 & 0 & 3 \\ 0 & 0 & 1 & 1 \\ 0 & 0 & 0 & 0 \end{bmatrix}$.

33. $\left\{ \begin{bmatrix} 1 \\ -2 \\ -3 \\ 1 \end{bmatrix}, \begin{bmatrix} 0 \\ 1 \\ 2 \\ 0 \end{bmatrix} \right\}$, because $\begin{bmatrix} 1 & 0 & 3 \\ -2 & 1 & -2 \\ -3 & 2 & -1 \\ 1 & 0 & 3 \end{bmatrix} \sim \begin{bmatrix} 1 & 0 & 3 \\ 0 & 1 & 4 \\ 0 & 0 & 0 \\ 0 & 0 & 0 \end{bmatrix}$.

35. $\left\{\begin{bmatrix}-1\\0\\1\end{bmatrix},\begin{bmatrix}1\\0\\0\end{bmatrix},\begin{bmatrix}0\\1\\0\end{bmatrix}\right\}$, because $\begin{bmatrix}-1&1&0&0\\0&0&1&0\\1&0&0&1\end{bmatrix}\sim\begin{bmatrix}-1&1&0&0\\0&1&0&1\\0&0&1&0\end{bmatrix}$

37. $\left\{\begin{bmatrix}-1\\0\\1\\0\end{bmatrix},\begin{bmatrix}1\\-1\\0\\0\end{bmatrix},\begin{bmatrix}1\\0\\0\\0\end{bmatrix},\begin{bmatrix}0\\0\\0\\1\end{bmatrix}\right\}$, because

$$\begin{bmatrix}-1&1&1&0&0&0\\0&-1&0&1&0&0\\1&0&0&0&1&0\\0&0&0&0&0&1\end{bmatrix}\sim\begin{bmatrix}-1&1&1&0&0&0\\0&-1&0&1&0&0\\0&0&1&1&1&0\\0&0&0&0&0&1\end{bmatrix}$$

39. A basis is $\left\{\begin{bmatrix}1\\2\\2\\-1\end{bmatrix},\begin{bmatrix}0\\-1\\2\\3\end{bmatrix}\right\}$; hence, the rank is 2. This can be seen from

$$\begin{bmatrix}1&2&2&-1\\0&-1&2&3\\1&1&4&2\end{bmatrix}\sim\begin{bmatrix}1&2&2&-1\\0&-1&2&3\\0&0&0&0\end{bmatrix}$$

41. A basis is $\left\{\begin{bmatrix}1\\2\\2\end{bmatrix},\begin{bmatrix}0\\-1\\2\end{bmatrix},\begin{bmatrix}0\\0\\1\end{bmatrix}\right\}$, so the rank is 3, by

$$\begin{bmatrix}1&2&2\\0&-1&2\\1&1&4\\0&1&-1\\0&0&2\end{bmatrix}\sim\begin{bmatrix}1&2&2\\0&-1&2\\0&0&1\\0&0&0\\0&0&0\end{bmatrix}$$

43. $\left\{\begin{bmatrix}1\\0\end{bmatrix},\begin{bmatrix}0\\1\end{bmatrix}\right\}$, by $\begin{bmatrix}1&1\\2&3\\-1&-2\end{bmatrix}\sim\begin{bmatrix}1&0\\0&1\\0&0\end{bmatrix}$.

45. $\left\{\begin{bmatrix}1\\0\\-2\end{bmatrix},\begin{bmatrix}0\\1\\-4\end{bmatrix}\right\}$, by $\begin{bmatrix}-1&1&-2\\2&-1&0\\1&0&-2\\1&-1&2\end{bmatrix}\sim\begin{bmatrix}1&0&-2\\0&1&-4\\0&0&0\\0&0&0\end{bmatrix}$.

47. $B \sim \begin{bmatrix} 1 & 1 & 2 & 2 \\ 0 & 0 & -1 & 2 \\ 0 & 0 & 0 & 0 \end{bmatrix}$. Rank$(A) = 2$. $B^T \sim \begin{bmatrix} 1 & 0 & 0 \\ 0 & -1 & 1 \\ 0 & 0 & 0 \\ 0 & 0 & 0 \end{bmatrix}$. Rank$(B^T) = 2$.

49. The nullity of B is 2 (a null space basis: $\{(-6,0,2,1), (-1,1,0,0)\}$). The rank is 2. $4 = 2 + 2$ equals the number of columns.

51. $B \sim \begin{bmatrix} 1 & 1 & 2 & 2 \\ 0 & 0 & -1 & 2 \\ 0 & 0 & 0 & 0 \end{bmatrix}$, so the rank of B is 2. And $[B : \mathbf{b}] \sim \begin{bmatrix} 1 & 1 & 2 & 2 & 1 \\ 0 & 0 & -1 & 2 & 0 \\ 0 & 0 & 0 & 0 & 0 \end{bmatrix}$ so the rank of $[B : \mathbf{b}]$ is, again, 2.

53. Yes. A has 450 columns and nullity 50, so its rank is 400, by the rank theorem. Hence, the column space has dimension 400. Therefore, the column space is all of $\mathbf{R}^{400}$. So, every 400-vector $\mathbf{b}$ is spanned by the columns of A. Therefore, $A\mathbf{x} = \mathbf{b}$ is consistent for all 400-vectors $\mathbf{b}$.

55. Because the rank equals the number of pivots of the matrix statements 1–3 are clearly equivalent. Statements 3–5 are equivalent by Theorem 12, Section 3.4. The equivalence of 5–7 follows from the definition of the null space and the fact the $\{\mathbf{0}\}$ is the only subspace of dimension 0. The equivalence of 1, 8 and 9 follows from the definition of the rank and Theorems and 29. Lastly, statements 1 and 10 are equivalent by Corollary 30.

Exercises 4.7

---

1. (a) $\begin{bmatrix} 1 \\ 1 \\ 0 \end{bmatrix}$ (b) $\begin{bmatrix} 0 \\ 1 \\ 1 \end{bmatrix}$ (c) $\begin{bmatrix} 0 \\ 1 \\ 1 \end{bmatrix}$ (d) $\begin{bmatrix} 0 \\ 0 \\ 0 \end{bmatrix}$

3. $A^2 = \begin{bmatrix} 0 & 1 & 1 \\ 1 & 0 & 1 \\ 1 & 1 & 0 \end{bmatrix}$, $A^3 = \begin{bmatrix} 1 & 0 & 1 \\ 0 & 1 & 1 \\ 1 & 1 & 0 \end{bmatrix}$

5. No. For example, $\mathbf{u} + \mathbf{v} + \mathbf{w} = \mathbf{0}$ is a linear dependence relation.

7. $\{(1,1,1)\}$ is a basis of the null space of A over Z_2. The only elements of the null space over Z_2 are $(1,1,1)$ and $(0,0,0)$. This can be seen from the reduction over Z_2

$$[A:0] = \begin{bmatrix} 1 & 0 & 1 & 0 \\ 0 & 1 & 1 & 0 \\ 1 & 1 & 0 & 0 \end{bmatrix} \sim \begin{bmatrix} 1 & 0 & 1 & 0 \\ 0 & 1 & 1 & 0 \\ 0 & 1 & 1 & 0 \end{bmatrix} \sim \begin{bmatrix} 1 & 0 & 1 & 0 \\ 0 & 1 & 1 & 0 \\ 0 & 0 & 0 & 0 \end{bmatrix}$$

So, $x_1 = r$, $x_2 = r$, where $r \in Z_2$.

Over $\mathbf{R}$ the null space of A is $\{0\}$ (because $[A:0] \sim [I:0]$), so the only basis is the empty set.

9. The matrices commute, so $AB = BA$. Hence

$$\begin{aligned} (A+B)^2 &= (A+B)(A+B) \\ &= A^2 + AB + BA + B^2 \\ &= A^2 + 2AB + B^2 \\ &= A^2 + B^2 \end{aligned}$$

because $2AB$ is the zero matrix over Z_2.

11. 0101010, because over Z_2

$$\begin{bmatrix} 0 & 1 & 0 & 1 \end{bmatrix} G = \begin{bmatrix} 0 & 1 & 0 & 1 & 2 & 1 & 2 \end{bmatrix} = \begin{bmatrix} 0 & 1 & 0 & 1 & 0 & 1 & 0 \end{bmatrix}$$

13. (a) $(1,1,1,1)$, because $H(1,1,1,1,0,1,1) = (3,4,3) = (1,0,1)$ which only affects the parity portion of the message.

(b) $(0,1,1,1)$, because $H(1,1,1,1,1,0,0) = (2,2,3) = (0,0,1)$ which changes the first entry.

15. (a) $(0,1,1,0)$, because $H(0,1,1,0,0,0,1) = (1,3,2) = (1,1,0)$ which only affects the parity portion of the message.

(b) $(0,1,1,0)$, because $H(0,1,0,0,0,1,1) = (2,3,1) = (0,1,1)$ which changes the third entry.

17. (a) $(1,1,1,0)$, because $H(1,1,1,0,0,1,0) = (1,3,2) = (1,1,0)$ which only affects the parity portion of the message.

(b) $(1,1,1,0)$, because $H(1,1,1,0,0,0,0) = (0,2,2) = (0,0,0)$ which does not change the message.

19. We have

$$d(\mathbf{u}, \mathbf{v}) = w(\mathbf{u} - \mathbf{v}) = w((\mathbf{u} - \mathbf{v}) - \mathbf{0}) = d(\mathbf{u} - \mathbf{v}, \mathbf{0}) = d(\mathbf{0}, \mathbf{u} - \mathbf{v})$$

21. If $\mathbf{u}$ and $\mathbf{v}$ are in N_H, then so is $\mathbf{u} - \mathbf{v}$, because N_H is a subspace. Hence, $w(\mathbf{u} - \mathbf{v}) \geq 3$, by Exercise 20 for distinct $\mathbf{u}$ and $\mathbf{v}$. Therefore, $d(\mathbf{u}, \mathbf{v}) \geq 3$ because $d(\mathbf{u}, \mathbf{v}) = w(\mathbf{u} - \mathbf{v})$.

Exercises 5.1

1. It is a transformation. Its domain is $\mathbf{R}^2$. Its codomain is $\mathbf{R}^2$.

3. It is a transformation. Its domain is $\mathbf{R} \times \mathbf{R}^+$ (where, $\mathbf{R}^+$ is the set of nonnegative real numbers). Its codomain is $\mathbf{R}^2$.

5. This is not a transformation, because the matrix product is undefined.

7. They are not equal, because they have different codomains.

9. (a) $n = 2$, $m = 2$ (b) $\mathbf{R}^2$ and $\mathbf{R}^2$

(c) $\{0\}$, because $T(x, y) = (x + y, x - y) = (0, 0)$ implies $x = 0$, $y = 0$.

(d) The range is $\mathbf{R}^2$, because $T(x, y) = (x+y, x-y) = (a, b)$ can be solved for all a and b, as we see from the two-pivoted coefficient matrix of the corresponding system:

$$\begin{bmatrix} 1 & 1 \\ 1 & -1 \end{bmatrix} \sim \begin{bmatrix} 1 & 1 \\ 0 & -2 \end{bmatrix}$$

11. (a) $n = 3$, $m = 2$ (b) $\mathbf{R}^3$ and $\mathbf{R}^2$

(c) The set is $\mathrm{Span}\{(1, -1, 1)\}$, because $T(x, y, z) = (x + y, x - z) = (0, 0)$ yields

$$\begin{bmatrix} 1 & 1 & 0 & 0 \\ 1 & 0 & -1 & 0 \end{bmatrix} \sim \begin{bmatrix} 1 & 0 & -1 & 0 \\ 0 & 1 & 1 & 0 \end{bmatrix}$$

Hence, $x = r$, $y = -r$, $z = r$, for any $r \in \mathbf{R}$.

(d) The range is $\mathbf{R}^2$, because $T(x, y, z) = (x + y, x - z) = (a, b)$ can be solved for all a and b, as we see from the two-pivoted coefficient matrix of the corresponding system

$$\begin{bmatrix} 1 & 1 & 0 \\ 1 & 0 & -1 \end{bmatrix} \sim \begin{bmatrix} 1 & 1 & 0 \\ 0 & -1 & -1 \end{bmatrix}$$

13. (a) $n = 3$, $m = 3$ (b) $\mathbf{R}^3$ and $\mathbf{R}^3$

(c) The set is $\mathrm{Span}\{(1, 0, 1), (0, 1, 0)\}$, because $T(x, y, z) = (x-z, -x+z, x-z) = (0, 0, 0)$ implies $x = z = r$, $y = s$, where r, s are any scalars.

(d) The range is Span$\{(1,-1,1)\}$, because $T(x,y,z)=(x-z,-x+z,x-z)=(a,b,c)$ implies

$$\begin{bmatrix} 1 & 0 & -1 & a \\ -1 & 0 & 1 & b \\ 1 & 0 & -1 & c \end{bmatrix} \sim \begin{bmatrix} 1 & 0 & -1 & a \\ 0 & 0 & 0 & b+a \\ 0 & 0 & 0 & c-a \end{bmatrix}$$

or $a=c=-b$.

15. (a) $n=4$, $m=2$ (b) $\mathbf{R}^4$ and $\mathbf{R}^2$

(c) The set is Span$\{(-1,-1,1,0),(2,0,0,1)\}$, because $\{(-1,-1,1,0),(2,0,0,1)\}$ is a basis for the null space of the matrix.

(d) The range is $\mathbf{R}^2$, because the matrix has two pivots.

17. $\left\{\begin{bmatrix} 1 \\ 0 \end{bmatrix}, \begin{bmatrix} 2 \\ 1 \end{bmatrix}\right\}$

19. A basis is $\left\{\begin{bmatrix} 1 \\ 0 \end{bmatrix}, \begin{bmatrix} 2 \\ 1 \end{bmatrix}\right\}$, because columns 1 and 2 are the only pivot columns of the matrix.

21. $\left\{\begin{bmatrix} 1 \\ -2 \\ 0 \end{bmatrix}, \begin{bmatrix} 0 \\ 0 \\ -1 \end{bmatrix}, \begin{bmatrix} -2 \\ 1 \\ 4 \end{bmatrix}\right\}$, because

$$\begin{bmatrix} 1 & 0 & -2 \\ -2 & 0 & 1 \\ 0 & -1 & 4 \end{bmatrix} \sim \begin{bmatrix} 1 & 0 & -2 \\ 0 & -1 & 4 \\ 0 & 0 & -3 \end{bmatrix}$$

so all three columns are pivot columns.

23. (a) $\begin{bmatrix} -1 \\ 0 \end{bmatrix}, \begin{bmatrix} 0 \\ -1 \end{bmatrix}, \begin{bmatrix} 1 \\ -2 \end{bmatrix}$. Rotation by 180°.

(b) $\begin{bmatrix} -1 \\ 0 \end{bmatrix}, \begin{bmatrix} 0 \\ 3 \end{bmatrix}, \begin{bmatrix} 1 \\ 6 \end{bmatrix}$. None.

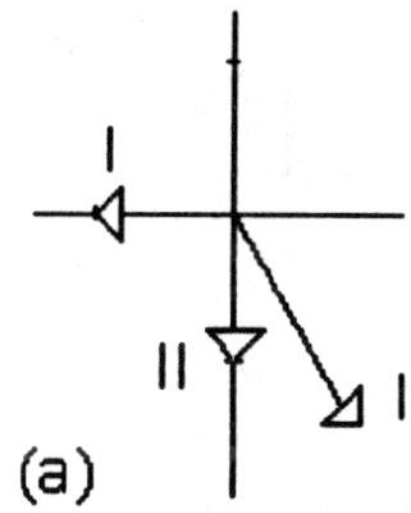

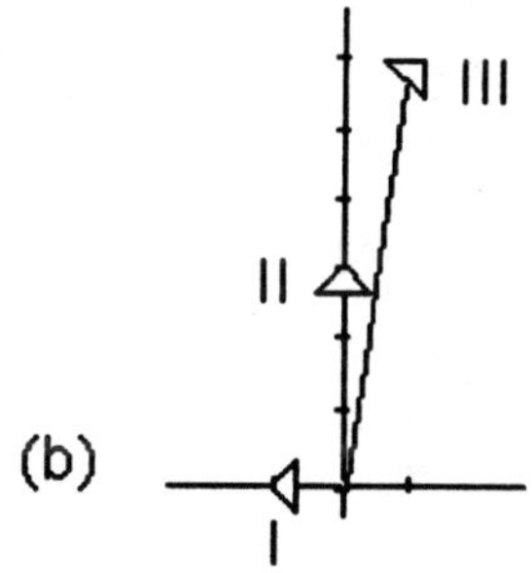

25. (a) $\begin{bmatrix} 1 \\ 0 \end{bmatrix}, \begin{bmatrix} -3 \\ 1 \end{bmatrix}, \begin{bmatrix} -7 \\ 2 \end{bmatrix}$. Shear by a factor of 3 along the opposite x-direction.

(b) $\begin{bmatrix} 1 \\ -3 \end{bmatrix}, \begin{bmatrix} -3 \\ 1 \end{bmatrix}, \begin{bmatrix} -7 \\ 5 \end{bmatrix}$. None.

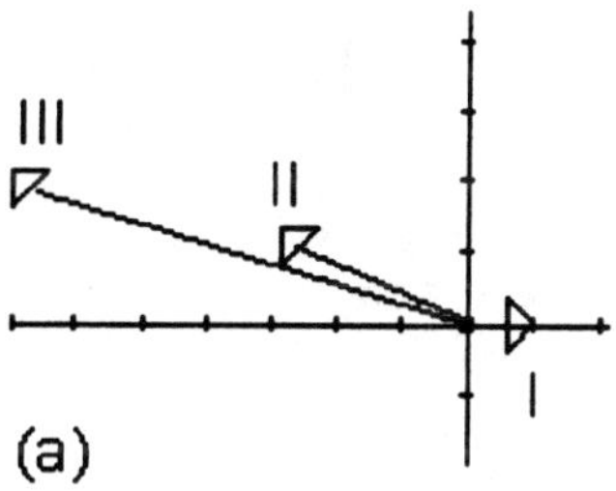

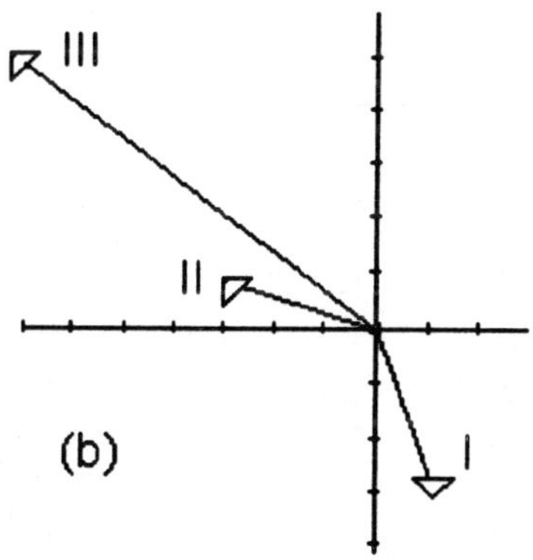

27. (a) $\begin{bmatrix} 1 \\ 3 \end{bmatrix}, \begin{bmatrix} 2 \\ 4 \end{bmatrix}, \begin{bmatrix} 3 \\ 5 \end{bmatrix}$. None.

(b) $\begin{bmatrix} 3 \\ 3 \end{bmatrix}, \begin{bmatrix} 3 \\ 3 \end{bmatrix}, \begin{bmatrix} 3 \\ 3 \end{bmatrix}$. None.

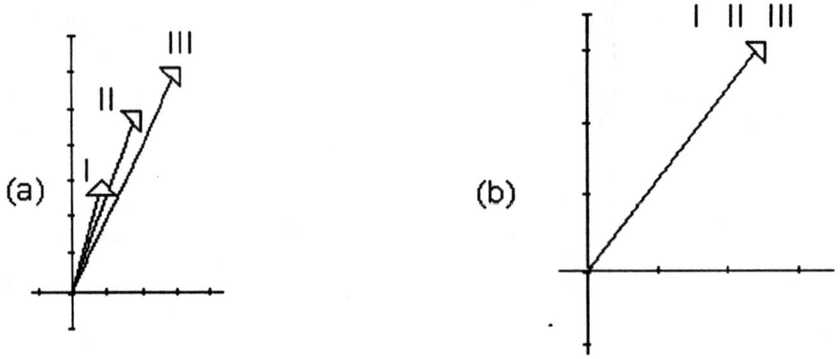

29. (a) $\begin{bmatrix} 1 \\ 0 \end{bmatrix}, \begin{bmatrix} 0 \\ 0 \end{bmatrix}, \begin{bmatrix} -1 \\ 0 \end{bmatrix}$. Projection onto the x-axis.

(b) $\begin{bmatrix} 3 \\ 0 \end{bmatrix}, \begin{bmatrix} 0 \\ 0 \end{bmatrix}, \begin{bmatrix} -3 \\ 0 \end{bmatrix}$. None.

31. $T(x, y) = (-y, -x),\ A = \begin{bmatrix} 0 & -1 \\ -1 & 0 \end{bmatrix}$

33. $R(x, y) = \begin{bmatrix} x\cos\theta + y\sin\theta \\ -x\sin\theta + y\cos\theta \end{bmatrix},\ A = \begin{bmatrix} \cos\theta & \sin\theta \\ -\sin\theta & \cos\theta \end{bmatrix}$

35. Any line in the plane is the set of vectors (points) of the form $\mathbf{x}_0 + r\mathbf{a}$, where $\mathbf{a}$ is a nonzero 2-vector (the direction), $\mathbf{x}_0$ is any 2-vector, and r varies through all real numbers. The image under T is

$$T(\mathbf{x}_0 + r\mathbf{a}) = T(\mathbf{x}_0) + T(r\mathbf{a}) = T(\mathbf{x}_0) + rT(\mathbf{a})$$

because T is a linear transformation. As r varies, $T(\mathbf{x}_0) + rT(\mathbf{a})$ defines a straight line, if $T(\mathbf{a})$ is nonzero, and the point $T(\mathbf{x}_0)$ if $T(\mathbf{a})$ is zero.

37. R^x_θ and R^y_θ with respective matrices:

$$\begin{bmatrix} 1 & 0 & 0 \\ 0 & \cos\theta & -\sin\theta \\ 0 & \sin\theta & \cos\theta \end{bmatrix}$$

and

$$\begin{bmatrix} \cos\theta & 0 & -\sin\theta \\ 0 & 1 & 0 \\ \sin\theta & 0 & \cos\theta \end{bmatrix}$$

Exercises 5.2

---

1. The domain is $\mathbf{R}^2$ and the codomain is $\mathbf{R}^2$. The transformation is linear. This is seen by an easy check of the definition, or by noting that T is a matrix transformation with standard matrix $\begin{bmatrix} 3 & 1 \\ 1 & -2 \end{bmatrix}$.

3. The domain is $\mathbf{R}^2$ and the codomain is $\mathbf{R}^2$. The transformation is nonlinear; because $(0,0)$ is mapped to $(-1,0)$.

5. The domain is $\mathbf{R}^2$ and the codomain is $\mathbf{R}^2$. The transformation is linear. In fact, it is a matrix transformation with standard matrix $\begin{bmatrix} 1 & 1 \\ 0 & 1 \end{bmatrix}$.

7. The domain is $\mathbf{R}^2$ and the codomain is $\mathbf{R}^2$. The transformation is linear, because

$$T(\mathbf{x}+\mathbf{y}) = 3(\mathbf{x}+\mathbf{y}) = 3\mathbf{x}+3\mathbf{y} = T(\mathbf{x})+T(\mathbf{y})$$

$$T(c\mathbf{x}) = 3(c\mathbf{x}) = c(3\mathbf{x}) = cT(\mathbf{x})$$

9. The domain is $\mathbf{R}^2$ and the codomain is $\mathbf{R}^3$. The transformation is linear, because it is a matrix transformation with standard matrix $\begin{bmatrix} 1 & 1 \\ 0 & 1 \\ 0 & 0 \end{bmatrix}$.

11. The domain is $\mathbf{R}^3$ and the codomain is $\mathbf{R}^2$. The transformation is linear, because it is a matrix transformation.

13. We have

$$\begin{aligned}
T\left(c_1\begin{bmatrix} x_1 \\ y_1 \\ z_1 \end{bmatrix} + c_2\begin{bmatrix} x_2 \\ y_2 \\ z_2 \end{bmatrix}\right) &= T\begin{bmatrix} c_1x_1 + c_2x_2 \\ c_1y_1 + c_2y_2 \\ c_1z_1 + c_2z_2 \end{bmatrix} \\
&= \begin{bmatrix} c_1x_1 + c_2x_2 + c_1y_1 + c_2y_2 + c_1z_1 + c_2z_2 \\ c_1x_1 + c_2x_2 - c_1y_1 - c_2y_2 - c_1z_1 - c_2z_2 \end{bmatrix} \\
&= c_1\begin{bmatrix} x_1 + y_1 + z_1 \\ x_1 - y_1 - z_1 \end{bmatrix} + c_2\begin{bmatrix} x_2 + y_2 + z_2 \\ x_2 - y_2 - z_2 \end{bmatrix} \\
&= c_1T\begin{bmatrix} x_1 \\ y_1 \\ z_1 \end{bmatrix} + c_2T\begin{bmatrix} x_2 \\ y_2 \\ z_2 \end{bmatrix}
\end{aligned}$$

15. Linear, because

$$\begin{aligned}
T\left((a_1 + b_1x) + (a_2 + b_2x)\right) &= T\left((a_1 + a_2) + (b_1 + b_2)\,x\right) \\
&= 3((a_1 + a_2) + (b_1 + b_2)) + ((a_1 + a_2) - 2(b_1 + b_2))x \\
&= (3a_1 + b_1) + (a_1 - 2b_1)x + (3a_2 + b_2) + (a_2 - 2b_2)x \\
&= T\left(a_1 + b_1x\right) + T\left(a_2 + b_2x\right)
\end{aligned}$$

and

$$\begin{aligned}
T\left(c\left(a + bx\right)\right) &= T\left(ca + cbx\right) \\
&= (3ca + cb) + (ca - 2cb)x \\
&= c\left((3a + b) + (a - 2b)x\right) \\
&= cT\left(a + bx\right)
\end{aligned}$$

17. Nonlinear, because $T(1 + x) = (1 - 5) + x = -4 + x$, whereas $T(1) + T(x) = (1 + x) + (-5 + x) = -4 + 2x$.

19. Nonlinear, because $T(0) = -2 \neq 0$.

21. Yes, because if $p_1 = a + bx + cx^2$ and $p_2 = A + Bx + Cx^2$, then

$$\begin{aligned} T(c_1p_1 + c_2p_2) &= T\left((c_1a + c_2A) + (c_1b + c_2B)x + (c_1c + c_2C)x^2\right) \\ &= \left((c_1a + c_2A) - (c_1b + c_2B)\right) + \left((c_1b + c_2B) + (c_1c + c_2C)\right)x \\ &= c_1\left((a-b) + (b+c)x\right) + c_2\left((A-B) + (B+C)x\right) \\ &= c_1T(p_1) + c_2T(p_2) \end{aligned}$$

23. No, because $T\begin{bmatrix} 0 \\ 0 \end{bmatrix} = \begin{bmatrix} 0 & 0 \\ 0 & 2 \end{bmatrix} \neq \mathbf{0}$.

25. Because the transformation is evaluated at $(1,1,1)$, $(-1,1,1)$, and $(1,-1,1)$ we have

$$\begin{bmatrix} 1 & -1 & 1 & 1 & 0 & 0 \\ 1 & 1 & -1 & 0 & 1 & 0 \\ 1 & 1 & 1 & 0 & 0 & 1 \end{bmatrix} \sim \begin{bmatrix} 1 & 0 & 0 & \frac{1}{2} & \frac{1}{2} & 0 \\ 0 & 1 & 0 & -\frac{1}{2} & 0 & \frac{1}{2} \\ 0 & 0 & 1 & 0 & -\frac{1}{2} & \frac{1}{2} \end{bmatrix}$$

so,

$$T\begin{bmatrix} 1 \\ 0 \\ 0 \end{bmatrix} = \frac{1}{2}\begin{bmatrix} 3 \\ -1 \end{bmatrix} - \frac{1}{2}\begin{bmatrix} 2 \\ -3 \end{bmatrix} = \begin{bmatrix} \frac{1}{2} \\ 1 \end{bmatrix}$$

$$T\begin{bmatrix} 0 \\ 1 \\ 0 \end{bmatrix} = \frac{1}{2}\begin{bmatrix} 3 \\ -1 \end{bmatrix} - \frac{1}{2}\begin{bmatrix} 2 \\ 1 \end{bmatrix} = \begin{bmatrix} \frac{1}{2} \\ -1 \end{bmatrix}$$

$$T\begin{bmatrix} 0 \\ 0 \\ 1 \end{bmatrix} = \frac{1}{2}\begin{bmatrix} 2 \\ -3 \end{bmatrix} + \frac{1}{2}\begin{bmatrix} 2 \\ 1 \end{bmatrix} = \begin{bmatrix} 2 \\ -1 \end{bmatrix}$$

Therefore,

$$T\begin{bmatrix} x \\ y \\ z \end{bmatrix} = \begin{bmatrix} \frac{1}{2} & \frac{1}{2} & 2 \\ 1 & -1 & -1 \end{bmatrix}\begin{bmatrix} x \\ y \\ z \end{bmatrix} = \begin{bmatrix} \frac{1}{2}x + \frac{1}{2}y + 2z \\ x - y - z \end{bmatrix}$$

and

$$T\begin{bmatrix} -10 \\ 15 \\ -25 \end{bmatrix} = \begin{bmatrix} \frac{1}{2} & \frac{1}{2} & 2 \\ 1 & -1 & -1 \end{bmatrix}\begin{bmatrix} -10 \\ 15 \\ -25 \end{bmatrix} = \begin{bmatrix} -\frac{95}{2} \\ 0 \end{bmatrix}$$

27. We have

$$1 = \tfrac{1}{2}\left(1 + x - x^2\right) + \tfrac{1}{2}\left(1 - x + x^2\right)$$
$$x = \tfrac{1}{2}\left(1 + x + x^2\right) - \tfrac{1}{2}\left(1 - x + x^2\right)$$
$$x^2 = \tfrac{1}{2}\left(1 + x + x^2\right) - \tfrac{1}{2}\left(1 + x - x^2\right)$$

Hence,

$$T(1) = \tfrac{1}{2}(-3 + 2x) + \tfrac{1}{2}(1 + 2x) = -1 + 2x$$
$$T(x) = \tfrac{1}{2}(-1 + 3x) - \tfrac{1}{2}(1 + 2x) = -1 + \tfrac{1}{2}x$$
$$T\left(x^2\right) = \tfrac{1}{2}(-1 + 3x) - \tfrac{1}{2}(-3 + 2x) = 1 + \tfrac{1}{2}x$$

Therefore,

$$\begin{aligned} T\left(a + bx + cx^2\right) &= a(-1 + 2x) + b\left(-1 + \tfrac{1}{2}x\right) + c\left(1 + \tfrac{1}{2}x\right) \\ &= -a - b + c + \left(2a + \tfrac{1}{2}b + \tfrac{1}{2}c\right)x \end{aligned}$$

and $T\left(-25 + 15x - 10x^2\right) = -\dfrac{95}{2}x.$

29. T is linear by Theorem 3, because

$$\begin{aligned} T(c_1X + c_2Y) &= C^{-1}(c_1X + c_2Y)C \\ &= C^{-1}(c_1XC + c_2YC) \\ &= C^{-1}(c_1XC) + C^{-1}(c_2YC) \\ &= c_1(C^{-1}XC) + c_2(C^{-1}YC) \\ &= c_1T(X) + c_2T(Y) \end{aligned}$$

31. Because $(-2, 2)$ is in the span of $\{(1, -1)\}$, T cannot be specified for any vector not in that span. Examples of different linear transformations such that $T(1, -1) = (3, -1)$ (thus, $T(-2, 2) = (-6, 2)$) are

$$T(x, y) = (-6x - 9y, 2x + 3y)$$
$$T(x, y) = (-3x - 6y, x + 2y)$$
$$T(x, y) = \left(-\tfrac{3}{2}x - \tfrac{9}{2}y, \tfrac{1}{2}x + \tfrac{3}{2}y\right)$$

33. $T(x, y) = (xy, x)$

35. $T(x, y) = (x, x)$, $\mathbf{v}_1 = (1, 0)$, $\mathbf{v}_2 = (1, 1)$

37. Any line in the plane is the set of vectors (points) of the form $\mathbf{x}_0 + r\mathbf{a}$ where $\mathbf{a}$ is a nonzero 2-vector (the direction), $\mathbf{x}_0$ is any 2-vector, and r varies through all real numbers. The image under T is

$$T(\mathbf{x}_0 + r\mathbf{a}) = T(\mathbf{x}_0) + T(r\mathbf{a}) = T(\mathbf{x}_0) + rT(\mathbf{a})$$

because T is linear. As r varies, $T(\mathbf{x}_0) + rT(\mathbf{a})$ defines a straight line if $T(\mathbf{a})$ is nonzero, and the point $T(\mathbf{x}_0)$ if $T(\mathbf{a})$ is zero.

39. It is nonlinear because $T(\mathbf{0}) = A\mathbf{0} + \mathbf{b} = \mathbf{b} \neq \mathbf{0}$.

41. $A = \begin{bmatrix} 1 & -1 \\ 0 & 1 \\ 1 & 0 \end{bmatrix}$

43. $A = \begin{bmatrix} 0 & 0 \\ 0 & 0 \\ 0 & 0 \end{bmatrix}$

45. $A = \begin{bmatrix} -1 & 0 \\ 0 & 1 \end{bmatrix}$

47. $A = \begin{bmatrix} \cos\theta & -\sin\theta \\ \sin\theta & \cos\theta \end{bmatrix}$

49. They are of the form $T(x) = cx$ for some fixed scalar c. Because, if $T(1) = c$ then $T(x) = T(1 \cdot x) = xT(1) = cx$.

Exercises 5.3

---

1. The reduction $\begin{bmatrix} 1 & -1 & 0 \\ 0 & 1 & -1 \\ -1 & 0 & 1 \end{bmatrix} \sim \begin{bmatrix} 1 & 0 & -1 \\ 0 & 1 & -1 \\ 0 & 0 & 0 \end{bmatrix}$ implies

kernel basis:	$\{(1, 1, 1)\}$
range basis:	$\{(1, 0, -1), (-1, 1, 0)\}$
nullity:	1
rank:	2
dimension theorem:	$1 + 2 = \dim(\mathbf{R}^3)$

3. The reduction $\begin{bmatrix} 1 & -2 \\ 1 & -2 \\ 2 & -4 \\ 0 & 0 \end{bmatrix} \sim \begin{bmatrix} 1 & -2 \\ 0 & 0 \\ 0 & 0 \\ 0 & 0 \end{bmatrix}$ implies

kernel basis:	$\{(2,1)\}$
range basis:	$\{(1,1,2,0)\}$
nullity:	1
rank:	1
dimension theorem:	$1+1=\dim(\mathbf{R}^2)$

5. Because $T(\mathbf{v}) = -2\mathbf{v} = \mathbf{0}$ implies $\mathbf{v} = \mathbf{0}$, and $T(-\mathbf{u}/2) = \mathbf{u}$ for any 3-vector $\mathbf{u}$, we have

kernel basis:	empty set
range basis:	$\{(1,0,0),(0,1,0),(0,0,1)\}$
nullity:	0
rank:	3
dimension theorem:	$0+3=\dim(\mathbf{R}^3)$

7. We have

kernel basis:	empty set
range basis:	$\{(1,0),(0,1)\}$
nullity:	0
rank:	2
dimension theorem:	$0+2=\dim(\mathbf{R}^2)$

9. Both the kernel and range can be computed from the reduction

$$\begin{bmatrix} 1 & -1 & 0 & 0 \\ 0 & 1 & 0 & -1 \\ 0 & 0 & 1 & 1 \end{bmatrix} \sim \begin{bmatrix} 1 & 0 & 0 & -1 \\ 0 & 1 & 0 & -1 \\ 0 & 0 & 1 & 1 \end{bmatrix}$$

Hence,

kernel basis:	$\{1+x-x^2+x^3\}$
range basis:	$\{1,-1+x,x^2\}$
nullity:	1
rank:	3
dimension theorem:	$1+3=\dim(P_3)$

11. The transformation is $T(a+bx+cx^2) = -c+ax+bx^2$. The reduction $\begin{bmatrix} 0 & 0 & -1 \\ 1 & 0 & 0 \\ 0 & 1 & 0 \end{bmatrix} \sim$ I_3 shows that the only kernel basis is the empty set. A range basis is $\{\mathbf{e}_1, \mathbf{e}_2, \mathbf{e}_3\} \subset \mathbf{R}^3$.

13. The transformation is given by $T(a + bx + cx^2) = c$, so a kernel basis is $\{1, x\}$ and a range basis is $\{1\}$.

15. Let X be a 3×3 matrix with entries $x_1, \ldots, x_9$. Then X is in the kernel of T if and only if

$$\begin{bmatrix} 0 & 0 & 0 \\ 0 & 0 & 0 \\ 0 & 0 & 0 \end{bmatrix} = \begin{bmatrix} 1 & 1 & 1 \\ 1 & 0 & 1 \\ 0 & 0 & 0 \end{bmatrix} \begin{bmatrix} x_1 & x_2 & x_3 \\ x_4 & x_5 & x_6 \\ x_7 & x_8 & x_9 \end{bmatrix}$$

$$= \begin{bmatrix} x_1 + x_4 + x_7 & x_2 + x_5 + x_8 & x_3 + x_6 + x_9 \\ x_1 + x_7 & x_2 + x_8 & x_3 + x_9 \\ 0 & 0 & 0 \end{bmatrix}$$

So, $x_1 + x_7 = 0$ and $x_1 + x_4 + x_7 = 0$, hence, $x_4 = 0$. Likewise, $x_5 = x_6 = 0$. Therefore, if $x_7 = r$, $x_8 = s$, $x_9 = t$, then $x_1 = -r$, $x_2 = -s$, $x_3 = -t$. We conclude that

$$\left\{ \begin{bmatrix} -1 & 0 & 0 \\ 0 & 0 & 0 \\ 1 & 0 & 0 \end{bmatrix}, \begin{bmatrix} 0 & -1 & 0 \\ 0 & 0 & 0 \\ 0 & 1 & 0 \end{bmatrix}, \begin{bmatrix} 0 & 0 & -1 \\ 0 & 0 & 0 \\ 0 & 0 & 1 \end{bmatrix} \right\}$$

is a basis for the kernel of T. Hence, $\dim(\text{Kernel}(T)) = 3$.

17. Let $X = \begin{bmatrix} x_1 & x_2 \\ x_3 & x_4 \end{bmatrix}$. X is in the kernel of T if and only if

$$\begin{bmatrix} 0 & 0 \\ 0 & 0 \end{bmatrix} = \begin{bmatrix} 2 & 0 \\ 0 & 3 \end{bmatrix} \begin{bmatrix} x_1 & x_2 \\ x_3 & x_4 \end{bmatrix} - \begin{bmatrix} x_1 & x_2 \\ x_3 & x_4 \end{bmatrix} \begin{bmatrix} 2 & 0 \\ 0 & 3 \end{bmatrix}$$

$$= \begin{bmatrix} 0 & -x_2 \\ x_3 & 0 \end{bmatrix}$$

So, $x_1 = r$, $x_2 = 0$, $x_3 = 0$, $x_4 = s$. Hence, a basis for the kernel is

$$\left\{ \begin{bmatrix} 1 & 0 \\ 0 & 0 \end{bmatrix}, \begin{bmatrix} 0 & 0 \\ 0 & 1 \end{bmatrix} \right\}$$

Thus, the nullity is 2; therefore, the rank is 2, by the dimension theorem.

19. Let $X = \begin{bmatrix} x_1 & x_2 \\ x_3 & x_4 \end{bmatrix}$. X is in the kernel of T if and only if

$$\begin{bmatrix} 0 & 0 \\ 0 & 0 \end{bmatrix} = \begin{bmatrix} 0 & 0 \\ 0 & 1 \end{bmatrix}\begin{bmatrix} x_1 & x_2 \\ x_3 & x_4 \end{bmatrix} - \begin{bmatrix} x_1 & x_2 \\ x_3 & x_4 \end{bmatrix}\begin{bmatrix} 0 & 0 \\ 0 & 1 \end{bmatrix}$$

$$= \begin{bmatrix} 0 & -x_2 \\ x_3 & 0 \end{bmatrix}.$$

So, $x_1 = r$, $x_2 = 0$, $x_3 = 0$, $x_4 = s$. Hence, a basis for the kernel is

$$\left\{ \begin{bmatrix} 1 & 0 \\ 0 & 0 \end{bmatrix}, \begin{bmatrix} 0 & 0 \\ 0 & 1 \end{bmatrix} \right\}$$

Thus, the nullity is 2; therefore, the rank is 2, by the dimension theorem.

21. The system $2x + y = 0$, $-3x + 4y = 0$ has only the trivial solution so, T is one-to-one by Theorem 9.

23. The system $2x - y = 0$, $x - y = 0$, $-x + y = 0$, $x - 2y = 0$ has only the trivial solution, so T is one-to-one by Theorem 9.

25. The system $2a - b = 0$, $-b + c = 0$, $-3a + c = 0$ has only the trivial solution, so T is one-to-one by Theorem 9.

27. Because the nullity is zero, the rank is 2, so the range is $\mathbf{R}^2$, hence, T is onto.

29. T is onto because the system $x - y + z = a$, $-x + y + z = b$ can be solved for any choices of a, b. For example, $x = \frac{1}{2}a - \frac{1}{2}b$, $y = 0$, $z = \frac{1}{2}a + \frac{1}{2}b$ is a solution, so for these choices, (x, y, z) maps to (a, b).

31. T is onto, because any the system $a + b = A$, $a + c = B$ can be solved for any A, B. For example, $a = B$, $b = A - B$, $c = 0$ is a solution, so for these choices, $a + bx + cx^2$ maps to $A + Bx$.

33. The kernel is zero, because $a - 2b = 0$, $-2a + b = 0$ has only the trivial solution. So, T is one-to-one. Hence, it is an isomorphism by Theorem 11.

35. The kernel is zero, because $x - y + z = 0$, $-x + y + z = 0$, $-y + z = 0$ has only the trivial solution. So, T is one-to-one. Thus, T is an isomorphism by Theorem 11.

37. The kernel is zero, because $c = 0$, $b = 0$, $a - b = 0$ has only the trivial solution. So, T is one-to-one. Hence, it is an isomorphism by Theorem 11.

39. Not isomorphism

41. Isomorphism

43. Not isomorphism

45. By Theorem 11 it suffices to show that T is one-to-one. By Theorem 9 it suffices to show that the kernel is zero. Let

$$T(\mathbf{v}) = [\mathbf{v}]_{\mathcal{B}} = \mathbf{0}$$

Then the coefficients of $\mathbf{v}$ as a linear combination of the vectors of $\mathcal{B}$ are zero. Hence, $\mathbf{v} = \mathbf{0}$. So, the kernel of T is zero. Therefore, T is an isomorphism.

47. If T is one-to-one, then $\dim(V) \leq \dim(W)$. If $\mathcal{B}$ is a basis of V, then $T(\mathcal{B})$ is linearly independent in W, by Theorem 10. If $\mathcal{B}$ has n elements, so does $T(\mathcal{B})$. But then $n \leq \dim(W)$, by the exchange theorem, because W contains at least n linearly independent elements.

49. If T is onto, then $\dim(V) \geq \dim(W)$. By the preceding exercise if $\mathcal{B}$ is a basis of V, then $T(\mathcal{B})$ spans W. So the number of elements of $T(\mathcal{B})$ is at least the dimension of W, by the exchange theorem. But this number is at most the dimension of V. Hence, $\dim(V) \geq \dim(W)$.

Exercises 5.4

1. The matrix is $\begin{bmatrix} 1 & -1 \\ 0 & -2 \\ 2 & 2 \end{bmatrix}$. To see this we evaluate T at $\mathcal{B}$ and write the answer in terms of $\mathcal{B}'$

$$T(1+x) = x + x^2 = a\,(-x+x^2) + b\,(1+x) + cx$$
$$T(1-x) = -2 + x - x^2 = a\,(-x+x^2) + b\,(1+x) + cx$$

Each right-hand side equals $b + (-a+b+c)\,x + ax^2$. The unknown coefficients can be computed by the combined reduction

$$\begin{bmatrix} 0 & 1 & 0 & 0 & -2 \\ -1 & 1 & 1 & 1 & 1 \\ 1 & 0 & 0 & 1 & -1 \end{bmatrix} \sim \begin{bmatrix} 1 & 0 & 0 & 1 & -1 \\ 0 & 1 & 0 & 0 & -2 \\ 0 & 0 & 1 & 2 & 2 \end{bmatrix}$$

3. The matrix is $\begin{bmatrix} 3 & -1 & 0 \\ 1 & 0 & -1 \\ 0 & 0 & 2 \end{bmatrix}$. To see this we evaluate T at $\mathcal{B}'$ and write the answer in terms of $\mathcal{B}'$

$$T\,(-x+x^2) = 1 - 2x + 3x^2 = a\,(-x+x^2) + b\,(1+x) + cx$$
$$T\,(1+x) = x - x^2 = a\,(-x+x^2) + b\,(1+x) + cx$$
$$T(x) = -1 + x = a\,(-x+x^2) + b\,(1+x) + cx$$

Each right-hand side equals $b + (-a+b+c)\,x + ax^2$. The unknown coefficients can now be found by

$$\begin{bmatrix} 0 & 1 & 0 & 1 & 0 & -1 \\ -1 & 1 & 1 & -2 & 1 & 1 \\ 1 & 0 & 0 & 3 & -1 & 0 \end{bmatrix} \sim \begin{bmatrix} 1 & 0 & 0 & 3 & -1 & 0 \\ 0 & 1 & 0 & 1 & 0 & -1 \\ 0 & 0 & 1 & 0 & 0 & 2 \end{bmatrix}$$

5. We have

(a) $A = \begin{bmatrix} 0 & 0 & -2 \\ 0 & 1 & 0 \\ 0 & 0 & 0 \end{bmatrix}$, as seeing from evaluating: $T(1) = 0, T(x) = x, T(x^2) = -2$.

(b) $A' = \begin{bmatrix} 0 & 0 & 0 \\ -2 & 0 & 0 \\ 1 & 1 & 1 \end{bmatrix}$, because

$$T(-x+x^2) = -2 - x = a(-x+x^2) + b(1+x) + cx$$
$$T(1+x) = x = a(-x+x^2) + b(1+x) + cx$$
$$T(x) = x = a(-x+x^2) + b(1+x) + cx$$

with each right-hand side being equal to $b + (-a+b+c)x + ax^2$ and the unknown coefficients computed from

$$\begin{bmatrix} 0 & 1 & 0 & -2 & 0 & 0 \\ -1 & 1 & 1 & -1 & 1 & 1 \\ 1 & 0 & 0 & 0 & 0 & 0 \end{bmatrix} \sim \begin{bmatrix} 1 & 0 & 0 & 0 & 0 & 0 \\ 0 & 1 & 0 & -2 & 0 & 0 \\ 0 & 0 & 1 & 1 & 1 & 1 \end{bmatrix}$$

(c) We evaluate directly to get $T(6x - 2x^2) = 4 + 6x$.

Using A we have

$$\begin{bmatrix} 0 & 0 & -2 \\ 0 & 1 & 0 \\ 0 & 0 & 0 \end{bmatrix} \begin{bmatrix} 0 \\ 6 \\ -2 \end{bmatrix} = \begin{bmatrix} 4 \\ 6 \\ 0 \end{bmatrix}$$

So, $4 + 6x$ in terms of the standard basis.

Using A' we have: First we write $6x - 2x^2$ in terms of $\mathcal{B}'$

$$6x - 2x^2 = -2(-x+x^2) + 0(1+x) + 4(x)$$

then we compute the matrix product

$$\begin{bmatrix} 0 & 0 & 0 \\ -2 & 0 & 0 \\ 1 & 1 & 1 \end{bmatrix} \begin{bmatrix} -2 \\ 0 \\ 4 \end{bmatrix} = \begin{bmatrix} 0 \\ 4 \\ 2 \end{bmatrix}$$

to get the coordinates of the image with respect to $\mathcal{B}'$. So,

$$T(6x - 2x^2) = 0(-x+x^2) + 4(1+x) + 2(x) = 4 + 6x$$

7. (a) $A = \begin{bmatrix} -1 & -1 \\ 1 & -1 \\ -3 & 1 \end{bmatrix}$. This is seen from the evaluations

$$T(1+x) = 1 - x - x^2 = a(-x+x^2) + b(1+x) + cx$$
$$T(-1+x) = -1 + x - x^2 = a(-x+x^2) + b(1+x) + cx$$

(with each right-hand side being equal to $b + (-a+b+c)\,x + ax^2$) and the reduction

$$\begin{bmatrix} 0 & 1 & 0 & 1 & -1 \\ -1 & 1 & 1 & -1 & 1 \\ 1 & 0 & 0 & -1 & -1 \end{bmatrix} \sim \begin{bmatrix} 1 & 0 & 0 & -1 & -1 \\ 0 & 1 & 0 & 1 & -1 \\ 0 & 0 & 1 & -3 & 1 \end{bmatrix}$$

(b) Direct evaluation yields $T(5-2x) = 5 - 5x + 2x^2$.

Using A we get

$$5 - 2x = \frac{3}{2}(1+x) - \frac{7}{2}(-1+x)$$

So

$$\begin{bmatrix} -1 & -1 \\ 1 & -1 \\ -3 & 1 \end{bmatrix} \begin{bmatrix} \frac{3}{2} \\ -\frac{7}{2} \end{bmatrix} = \begin{bmatrix} 2 \\ 5 \\ -8 \end{bmatrix}$$

Hence,

$$T(5-2x) = 2(-x+x^2) + 5(1+x) - 8x = 5 - 5x + 2x^2$$

9. The coordinate vector $\begin{bmatrix} 1 & -2 \\ -4 & 3 \end{bmatrix}$ with respect to $\mathcal{R}$ is $(2, 0, -4, -1)$. Hence,

$$\begin{bmatrix} 1 & 3 & 0 & 2 \\ 0 & 1 & 0 & 0 \\ 0 & -2 & 1 & -1 \\ 1 & 0 & 0 & 0 \end{bmatrix} \begin{bmatrix} 2 \\ 0 \\ -4 \\ -1 \end{bmatrix} = \begin{bmatrix} 0 \\ 0 \\ -3 \\ 2 \end{bmatrix}$$

and

$$T\begin{bmatrix} 1 & -2 \\ -4 & 3 \end{bmatrix} = -3\begin{bmatrix} 0 & 0 \\ 1 & -1 \end{bmatrix} + 2\begin{bmatrix} 1 & 0 \\ 0 & 1 \end{bmatrix} = \begin{bmatrix} 2 & 0 \\ -3 & 5 \end{bmatrix}$$

11. Suppose A and A' be two transition matrices for the linear transformation $T : V \to W$ with respect to bases $\mathcal{B}$ and $\mathcal{B}'$ of V and W, respectively. We shall show that $A = A'$. First we have that for all $\mathbf{v}$ in V

$$[T(\mathbf{v})]_{\mathcal{B}'} = A\,[\mathbf{v}]_{\mathcal{B}} = A'\,[\mathbf{v}]_{\mathcal{B}}$$

If $\mathcal{B} = \{\mathbf{v}_1, \ldots, \mathbf{v}_n\}$, then $[\mathbf{v}_i]_{\mathcal{B}} = \mathbf{e}_i$ for $i = 1, \ldots, n$. Therefore,

$$A\mathbf{e}_i = A[\mathbf{v}_i]_{\mathcal{B}} = A'[\mathbf{v}_i]_{\mathcal{B}} = A'\mathbf{e}_i$$

for $i = 1, \ldots, n$. Hence, A and A' have the same columns, so they are equal.

13. We have $m = 2, n = 4$ and

$$T(a + bx + cx^2 + dx^3) = (-9a - b + 8c + 5d) + (11a + 3b - 10c + d)x$$

To see this we write $a + bx + cx^2 + dx^3$ in terms of $\mathcal{B}''$

$$\begin{aligned} a + bx + cx^2 + dx^3 &= A(-x + x^3) + B(1 + x^2) + Cx + D(-1 + x) \\ &= (-D + B) + (-A + C + D)x + Bx^2 + Ax^3 \end{aligned}$$

and solve the corresponding system to get the components of the coordinate vector $A = d$, $B = c$, $C = b + d + a - c$, $D = -a + c$. Then we multiply by the matrix to get the coordinate vector of the image

$$\begin{bmatrix} 2 & 0 & 1 & 0 \\ -4 & 1 & 2 & -8 \end{bmatrix} \begin{bmatrix} d \\ c \\ a - c + b + d \\ -a + c \end{bmatrix} = \begin{bmatrix} a + b - c + 3d \\ 10a + 2b - 9c - 2d \end{bmatrix}$$

Finally, we use $\mathcal{B}$ to get

$$\begin{aligned} T(a + bx + cx^2 + dx^3) &= (a + b - c + 3d)(1 + x) + (10a + 2b - 9c - 2d)(-1 + x) \\ &= (-9a - b + 8c + 5d) + (11a + 3b - 10c + d)x \end{aligned}$$

15. The matrix is

$$A = \begin{bmatrix} 1 & 0 & 0 & 0 \\ 1 & 0 & -1 & 1 \\ 1 & -1 & 0 & 1 \\ 0 & -1 & 0 & 0 \end{bmatrix}$$

and it is seen easily that it reduces to I_4. So A is invertible, thus T is an isomorphism by Theorem 16.

17. Let B be similar to A. Then there is invertible P such that $B = P^{-1}AP$. Hence, $A = PBP^{-1}$ or $A = (P^{-1})^{-1}BP^{-1}$. Hence, A is similar to B, because P^{-1} is invertible.

19. If A is invertible then $AB = A(BA)A^{-1}$, hence, AB is similar to BA. If B is invertible then $AB = B^{-1}(BA)B$, hence, again AB is similar to BA.

E x e r c i s e s 5.5

1. We have

$$(f+g)(x,y,z)=\begin{bmatrix} z \\ 3x-z \end{bmatrix} \qquad (f+g)(-1,2,0)=\begin{bmatrix} 0 \\ -3 \end{bmatrix}$$

$$(-4f)(x,y,z)=\begin{bmatrix} -4x+4y-4z \\ -4x-4y \end{bmatrix} \qquad (-4f)(-1,2,0)=\begin{bmatrix} 12 \\ -4 \end{bmatrix}$$

3. We have

$$(f+g)(x,y,z)=\begin{bmatrix} -x-y+z \\ 2x-3y-z \end{bmatrix} \qquad (f+g)(-1,2,0)=\begin{bmatrix} -1 \\ -8 \end{bmatrix}$$

$$(-4f)(x,y,z)=\begin{bmatrix} 4x+8y \\ -4x+4z \end{bmatrix} \qquad (-4f)(-1,2,0)=\begin{bmatrix} 12 \\ 4 \end{bmatrix}$$

5. If M(f) denotes the matrix of any function f, then

$$\mathrm{M}(f)+\mathrm{M}(g)=\begin{bmatrix} 1 & -1 & 1 \\ 1 & 1 & 0 \end{bmatrix}+\begin{bmatrix} -1 & 1 & 0 \\ 2 & -1 & -1 \end{bmatrix}=\begin{bmatrix} 0 & 0 & 1 \\ 3 & 0 & -1 \end{bmatrix}=\mathrm{M}(f+g)$$

$$-2\mathrm{M}(f)=\begin{bmatrix} -2 & 2 & -2 \\ -2 & -2 & 0 \end{bmatrix}=\mathrm{M}(-2f)$$

7. If M(f) denotes the matrix of any function f, then

$$\mathrm{M}(f)+\mathrm{M}(g)=\begin{bmatrix} -1 & -2 & 0 \\ 1 & 0 & -1 \end{bmatrix}+\begin{bmatrix} 0 & 1 & 1 \\ 1 & -3 & 0 \end{bmatrix}=\begin{bmatrix} -1 & -1 & 1 \\ 2 & -3 & -1 \end{bmatrix}=\mathrm{M}(f+g)$$

$$-2\mathrm{M}(f)=\begin{bmatrix} 2 & 4 & 0 \\ -2 & 0 & 2 \end{bmatrix}=\mathrm{M}(-2f)$$

9. $f \circ g(x, y) = (x + 4y, 3x - y)$, $f \circ g(-1, -3) = (-13, 0)$

11. If $\mathrm{M}(f)$ denotes the matrix of f, then

$$\mathrm{M}(f) \cdot \mathrm{M}(g) = \begin{bmatrix} 1 & -1 \\ 1 & 1 \end{bmatrix} \begin{bmatrix} -5 & 1 \\ 1 & 3 \end{bmatrix} = \begin{bmatrix} -6 & -2 \\ -4 & 4 \end{bmatrix} = \mathrm{M}(f \circ g)$$

13. If $\mathrm{M}(f)$ denotes the matrix of f, then

$$\mathrm{M}(f) \cdot \mathrm{M}(g) = \begin{bmatrix} -1 & -2 & 1 & 0 \\ 1 & 0 & 0 & -1 \end{bmatrix} \begin{bmatrix} 0 & 1 \\ 1 & -3 \\ 1 & -1 \\ 1 & 0 \end{bmatrix} = \begin{bmatrix} -1 & 4 \\ -1 & 1 \end{bmatrix} = \mathrm{M}(f \circ g)$$

15. (a) The codomain of g is $\mathbf{R}^3$, whereas the domain of f is $\mathbf{R}^2$. $g \circ f$ is undefined.

(b) The codomain of g is $\mathbf{R}^2$, whereas the domain of f is $\mathbf{R}^3$. $g \circ f$ is defined.

(c) The codomain of g is $\mathbf{R}^3$, whereas the domain of f is $\mathbf{R}^2$. $g \circ f$ is defined.

17. $f^3(x, y) = (-2x - 2y, 2x - 2y)$, $f^3(1, -1) = (0, 4)$

19. f is invertible, because its standard matrix $\begin{bmatrix} -1 & 1 & 1 \\ 1 & 0 & -2 \\ 2 & -1 & 0 \end{bmatrix}$ is invertible.

21. f is invertible, because its matrix is invertible with inverse $\begin{bmatrix} -\frac{1}{2} & 0 & \frac{1}{4} \\ 0 & -1 & 1 \\ 0 & 0 & -\frac{1}{2} \end{bmatrix}$.

Hence,

$$f^{-1}(x, y, z) = \left(-\frac{1}{2}x + \frac{1}{4}z, -y + z, -\frac{1}{2}z\right)$$

23. f is invertible, because its matrix is invertible with inverse $\begin{bmatrix} \frac{5}{3} & -\frac{7}{6} & \frac{1}{6} \\ -1 & 1 & 0 \\ -\frac{1}{3} & -\frac{1}{6} & \frac{1}{6} \end{bmatrix}$.

Hence,

$$f^{-1}(x, y, z) = \left(\frac{5}{3}x - \frac{7}{6}y + \frac{1}{6}z, -x + y, -\frac{1}{3}x - \frac{1}{6}y + \frac{1}{6}z\right)$$

25. The matrix is invertible, so f is invertible. f^{-1} is multiplication on the left by the inverse of the given matrix. Therefore, $f^{-1} : \mathbf{R}^4 \to \mathbf{R}^4$ is given by

$$f^{-1}(\mathbf{x}) = \begin{bmatrix} -1 & 0 & 1 & 2 \\ 0 & 0 & 1 & 1 \\ -1 & 1 & 1 & 2 \\ -1 & 1 & 1 & 1 \end{bmatrix} \mathbf{x}$$

27. 1. We have

$$\begin{aligned} [(f+g)+h](x) &= (f+g)(x) + h(x) \\ &= f(x) + g(x) + h(x) \\ &= f(x) + (g+h)(x) \\ &= [f + (g+h)](x) \end{aligned}$$

2. $(f+g)(x) = f(x) + g(x) = g(x) + f(x) = (g+f)(x)$

3. This is proved by property 2 and by

$$(f+\mathbf{0})(x) = f(x) + \mathbf{0}(x) = f(x) + 0 = f(x)$$

4. This is proved by property 2 and by

$$(f+(-f))(x) = f(x) + (-f)(x) = f(x) - f(x) = 0$$

5. We have

$$\begin{aligned} c(f+g)(x) &= c(f(x) + g(x)) \\ &= cf(x) + cg(x) \\ &= (cf)(x) + (cg)(x) \end{aligned}$$

6. $(a+b)f(x) = af(x) + bf(x) = (af)(x) + (bf)(x)$

7. $(ab)f(x) = abf(x) = a(bf)(x) = b(af)(x)$

8. $1f(x) = f(x)$

9. $0f(x) = 0$

29. 2. We have

$$\begin{aligned} [(f-g)(\mathbf{v})]_{\mathcal{B}'} &= [f(\mathbf{v}) - g(\mathbf{v})]_{\mathcal{B}'} \\ &= [f(\mathbf{v})]_{\mathcal{B}'} - [g(\mathbf{v})]_{\mathcal{B}'} \\ &= A[\mathbf{v}]_{\mathcal{B}} - B[\mathbf{v}]_{\mathcal{B}} \\ &= (A-B)[\mathbf{v}]_{\mathcal{B}} \end{aligned}$$

3. $[(-f)(\mathbf{v})]_{\mathcal{B}'} = [-f(\mathbf{v})]_{\mathcal{B}'} = -[f(\mathbf{v})]_{\mathcal{B}'} = -A[\mathbf{v}]_{\mathcal{B}} = (-A)[\mathbf{v}]_{\mathcal{B}}$
4. $[(cf)(\mathbf{v})]_{\mathcal{B}'} = [cf(\mathbf{v})]_{\mathcal{B}'} = c[f(\mathbf{v})]_{\mathcal{B}'} = cA[\mathbf{v}]_{\mathcal{B}} = (cA)[\mathbf{v}]_{\mathcal{B}}$
5. $[f \circ g(\mathbf{v})]_{\mathcal{B}'} = [f(g(\mathbf{v}))]_{\mathcal{B}'} = A[g(\mathbf{v})]_{\mathcal{B}''} = A(B\,[\mathbf{v}]_{\mathcal{B}}) = (AB)[\mathbf{v}]_{\mathcal{B}}$

31. We have the following.

1. For the equivalence of (a) and (b) we have: Let $f : V \to W$ have a right inverse $g : W \to V$. So, $f \circ g = I$. We shall prove that f is onto. Let $\mathbf{w}$ be any vector of W and let $\mathbf{v} = g(\mathbf{w})$. Then $f(\mathbf{v}) = f(g(\mathbf{w})) = f \circ g(\mathbf{w}) = \mathbf{w}$. So, f is onto.

   Conversely, if f is onto we define a function $g : W \to V$ as follows. Let $\{\mathbf{w}_1, \dots, \mathbf{w}_m\}$ be a basis of W. Because f is onto, there are vectors $\mathbf{v}_1, \dots, \mathbf{v}_m$ of V such that $f(\mathbf{v}_1) = \mathbf{w}_1, \dots, f(\mathbf{v}_m) = \mathbf{w}_m$. If $\mathbf{w}$ is any vector of W, then there are scalars $c_1\mathbf{w}_1 + \cdots + c_m\mathbf{w}_m$. We define g by

   $$g(\mathbf{w}) = c_1\mathbf{v}_1 + \cdots + c_m\mathbf{v}_m$$

   g is well defined by the uniqueness of the scalars. Lastly, g is linear, because, if $\mathbf{u}_1 = r_1\mathbf{w}_1 + \cdots + r_m\mathbf{w}_m$ and $\mathbf{u}_2 = s_1\mathbf{w}_1 + \cdots + s_m\mathbf{w}_m$, then

   $$\begin{aligned} k_1\mathbf{u}_1 + k_2\mathbf{u}_2 &= k_1(r_1\mathbf{w}_1 + \cdots + r_m\mathbf{w}_m) + k_2(s_1\mathbf{w}_1 + \cdots + s_m\mathbf{w}_m) \\ &= (k_1r_1 + k_2s_1)\,\mathbf{w}_1 + \cdots + (k_1r_m + k_2s_m)\,\mathbf{w}_m \end{aligned}$$

   Therefore,

   $$\begin{aligned} g(k_1\mathbf{u}_1 + k_2\mathbf{u}_2) &= (k_1r_1 + k_2s_1)\,\mathbf{v}_1 + \cdots + (k_1r_m + k_2s_m)\,\mathbf{v}_m \\ &= k_1(r_1\mathbf{v}_1 + \cdots + r_m\mathbf{v}_m) + k_2(s_1\mathbf{v}_1 + \cdots + s_m\mathbf{v}_m) \\ &= k_1g(\mathbf{u}_1) + k_2g(\mathbf{u}_2) \end{aligned}$$

2. For the equivalence of (a) and (c) we have: Suppose that f has a right inverse g. Let $\mathcal{B}$ and $\mathcal{B}'$ be bases of V and W and let A and B be the matrices of f and g with respect to $\mathcal{B}, \mathcal{B}'$ and $\mathcal{B}', \mathcal{B}$. Then the matrix of $f \circ g = I$ is the identity I and also the product AB, by Theorem 22. Hence, A has a right inverse, namely B.

   Conversely, let A be the matrix of f with respect to bases $\mathcal{B} = \{\mathbf{v}_1, \dots, \mathbf{v}_n\}$ and $\mathcal{B}' = \{\mathbf{w}_1, \dots, \mathbf{w}_m\}$ of V and W, respectively. Suppose that A has a right inverse, say B. Hence, $AB = I$. We define a transformation $g : W \to V$

as follows. Let $\mathbf{w} = c_1\mathbf{w}_1 + \cdots + c_m\mathbf{w}_m$. Then $[\mathbf{w}]_{\mathcal{B}'} = (c_1, \ldots, c_m)$. If $(k_1, \ldots, k_n) = B[\mathbf{w}]_{\mathcal{B}'}$, then we let

$$g(\mathbf{w}) = k_1\mathbf{v}_1 + \cdots + k_n\mathbf{v}_n$$

g is well defined and its matrix with respect to $\mathcal{B}'$ and $\mathcal{B}$ is B, by its definition. In addition, $f \circ g = I$ because,

$$[f \circ g(\mathbf{w})]_{\mathcal{B}'} = A[g(\mathbf{w})]_{\mathcal{B}} = AB[\mathbf{w}]_{\mathcal{B}'} = I[\mathbf{w}]_{\mathcal{B}'} = [\mathbf{w}]_{\mathcal{B}'}$$

Finally, g is linear. Indeed, $g(r_1\mathbf{u}_1 + r_2\mathbf{u}_2) = r_1 g(\mathbf{u}_1) + r_2 g(\mathbf{u}_2)$, because

$$\begin{aligned} [g(r_1\mathbf{u}_1 + r_2\mathbf{u}_2)]_{\mathcal{B}'} &= B[r_1\mathbf{u}_1 + r_2\mathbf{u}_2]_{\mathcal{B}'} \\ &= r_1 B[\mathbf{u}_1]_{\mathcal{B}'} + r_2 B[\mathbf{u}_2]_{\mathcal{B}'} \\ &= r_1[g(\mathbf{u}_1)]_{\mathcal{B}'} + r_2[g(\mathbf{u}_2)]_{\mathcal{B}'} \end{aligned}$$

33. All equivalences follow easily from the last two exercises and the following:

$$h = h \circ I = h \circ (f \circ g) = (h \circ f) \circ g = I \circ g = g$$

35. Because $f^{-1}\begin{bmatrix} x \\ y \end{bmatrix} = \begin{bmatrix} \frac{1}{5}x + \frac{2}{5}y \\ -\frac{2}{5}x + \frac{1}{5}y \end{bmatrix}$ and $f^{-2}\begin{bmatrix} x \\ y \end{bmatrix} = \begin{bmatrix} -\frac{3}{25}x + \frac{4}{25}y \\ -\frac{4}{25}x - \frac{3}{25}y \end{bmatrix}$, we have

$$f^{-2} \circ f^{-1}\begin{bmatrix} x \\ y \end{bmatrix} = \begin{bmatrix} -\frac{11}{125}x - \frac{2}{125}y \\ \frac{2}{125}x - \frac{11}{125}y \end{bmatrix}$$

which is also the formula for $f^{-3}(x, y)$.

Exercises 5.6

1. We have

$$T(\mathbf{0}) = \begin{bmatrix} 1 \\ -1 \end{bmatrix}, \quad T(\mathbf{e}_1) = \begin{bmatrix} 2 \\ -1 \end{bmatrix}, \quad T(\mathbf{e}_2) = \begin{bmatrix} 1 \\ 4 \end{bmatrix}$$

3. We have

$$T(\mathbf{0}) = \begin{bmatrix} 1 \\ -1 \\ 0 \end{bmatrix}, \quad T(\mathbf{e}_1) = \begin{bmatrix} 0 \\ 0 \\ 0 \end{bmatrix}, \quad T(\mathbf{e}_2) = \begin{bmatrix} 3 \\ 1 \\ -4 \end{bmatrix}, \quad T(\mathbf{e}_3) = \begin{bmatrix} 1 \\ -2 \\ 1 \end{bmatrix}$$

5. $\begin{bmatrix} 1 & -1 \\ -1 & 1 \end{bmatrix}\mathbf{x} + \begin{bmatrix} 0 \\ -1 \end{bmatrix}$

7. $\begin{bmatrix} -1 & 3 & 0 \\ 1 & 0 & -1 \\ 1 & -5 & 1 \end{bmatrix}\mathbf{x} + \begin{bmatrix} 1 \\ 0 \\ -1 \end{bmatrix}$

9. $A = \begin{bmatrix} 0 & 5 \\ 2 & -8 \end{bmatrix}$, $\mathbf{b} = \begin{bmatrix} -1 \\ 1 \end{bmatrix}$. This can be seen as follows. If $T(\mathbf{x}) = A\mathbf{x} + \mathbf{b}$, then $T(\mathbf{0}) = \mathbf{b} = \begin{bmatrix} -1 \\ 1 \end{bmatrix}$. Let $A = \begin{bmatrix} a & b \\ c & d \end{bmatrix}$. We have

$$A\mathbf{e}_1 + \mathbf{b} = \begin{bmatrix} a-1 \\ c+1 \end{bmatrix} = \begin{bmatrix} -1 \\ 3 \end{bmatrix} \quad \text{and} \quad A\mathbf{e}_2 + \mathbf{b} = \begin{bmatrix} b-1 \\ d+1 \end{bmatrix} = \begin{bmatrix} 4 \\ -7 \end{bmatrix}$$

Equating the entries of the vectors and solving for a, b, c, d yields $a = 0$, $b = 5$, $c = 2$, $d = -8$.

11. $A = \begin{bmatrix} 1 & -2 \\ 3 & 4 \end{bmatrix}$, $\mathbf{b} = \begin{bmatrix} -3 \\ 2 \end{bmatrix}$. This can be as follows. Let $T(\mathbf{x}) = \begin{bmatrix} a & b \\ c & d \end{bmatrix}\mathbf{x} + \begin{bmatrix} e \\ f \end{bmatrix}$. Then

$$T(\mathbf{e}_1) = \begin{bmatrix} a+e \\ c+f \end{bmatrix} = \begin{bmatrix} -2 \\ 5 \end{bmatrix}$$

$$T(\mathbf{e}_2) = \begin{bmatrix} b+e \\ d+f \end{bmatrix} = \begin{bmatrix} -5 \\ 6 \end{bmatrix}$$

$$T(\mathbf{e}_1 + \mathbf{e}_2) = \begin{bmatrix} a+b+e \\ c+d+f \end{bmatrix} = \begin{bmatrix} -4 \\ 9 \end{bmatrix}$$

Equating the corresponding entries of the equal vectors and solving the system yields $a = 1$, $b = -2$, $c = 3$, $d = 4$, $e = -3$, $f = 2$.

13. $A = \begin{bmatrix} 1 & -2 & 3 \\ -4 & 5 & -6 \end{bmatrix}$, $\mathbf{b} = \begin{bmatrix} 1 \\ -1 \end{bmatrix}$. This can be seen by letting $T(\mathbf{x}) = A\mathbf{x} + \mathbf{b}$ with $A = \begin{bmatrix} a_1 & a_2 & a_3 \\ a_4 & a_5 & a_6 \end{bmatrix}$ and $\mathbf{b} = \begin{bmatrix} b_1 \\ b_2 \end{bmatrix}$. The information on the values of T can be assembled into the following matrix equation

$$\begin{bmatrix} a_1 & a_2 & a_3 \\ a_4 & a_5 & a_6 \end{bmatrix} \begin{bmatrix} 1 & 0 & 0 & 1 \\ 0 & 1 & 0 & 1 \\ 0 & 0 & 1 & 1 \end{bmatrix} + \begin{bmatrix} b_1 \\ b_2 \end{bmatrix} = \begin{bmatrix} 2 & -1 & 4 & 3 \\ -5 & 4 & -7 & -6 \end{bmatrix}$$

Therefore,

$$\begin{bmatrix} a_1 + b_1 & a_2 + b_1 & a_3 + b_1 & a_1 + a_2 + a_3 + b_1 \\ a_4 + b_2 & a_5 + b_2 & a_6 + b_2 & a_4 + a_5 + a_6 + b_2 \end{bmatrix} = \begin{bmatrix} 2 & -1 & 4 & 3 \\ -5 & 4 & -7 & -6 \end{bmatrix}$$

Equating entries and solving the resulting system yields

$$a_1 = 1,\ a_2 = -2,\ a_3 = 3,\ a_4 = -4,\ a_5 = 5,\ a_6 = -6,\ b_1 = 1,\ b_2 = -1$$

15. Because $T(0) = \mathbf{b}$, $\mathbf{b}$ is determined by $T(0)$. Also, because $L(\mathbf{x}) = T(\mathbf{x}) - \mathbf{b} = A\mathbf{x}$ is a matrix transformation, it can be uniquely determined by $T(\mathbf{e}_1), \ldots, T(\mathbf{e}_n)$, so that T is uniquely determined by the values

$$T(\mathbf{e}_1),\ \ldots,\ T(\mathbf{e}_n),\ T(0)$$

17. The two sets are equal.

Exercises 6.1

1. (a) $3(7) - 6(4) = -3$ (b) $3(7) - 4(6) = -3$

3. (a) $1\begin{vmatrix} 5 & 0 \\ 0 & 4 \end{vmatrix} - 0\begin{vmatrix} 0 & 0 \\ 3 & 4 \end{vmatrix} + 2\begin{vmatrix} 0 & 5 \\ 3 & 0 \end{vmatrix} = -10$ (b) $5\begin{vmatrix} 1 & 2 \\ 3 & 4 \end{vmatrix} = -10$

5. (a) $2\begin{vmatrix} 11 & 1 \\ -2 & 7 \end{vmatrix} = 158$ (b) $-3\begin{vmatrix} 0 & -2 \\ 3 & 0 \end{vmatrix} - 1\begin{vmatrix} 5 & 1 \\ 3 & 0 \end{vmatrix} = -15$

7. (a) $1\begin{vmatrix} 0 & 0 & 1 \\ 1 & 0 & 0 \\ 0 & 1 & 0 \end{vmatrix} = -1\begin{vmatrix} 0 & 1 \\ 1 & 0 \end{vmatrix} = 1$ (b) $1(2)(3)(0) = 0$

9. The determinant equals

$$1\begin{vmatrix} 1 & -1 & 2 & 0 \\ 0 & 4 & 2 & 0 \\ 0 & 5 & 0 & 0 \\ 4 & 6 & 7 & 8 \end{vmatrix} = 8\begin{vmatrix} 1 & -1 & 2 \\ 0 & 4 & 2 \\ 0 & 5 & 0 \end{vmatrix} = 8\begin{vmatrix} 4 & 2 \\ 5 & 0 \end{vmatrix} = -80$$

11. Cofactor expansion about the last row yields

$$b\begin{vmatrix} a & a \\ a & a \end{vmatrix} - c\begin{vmatrix} a & a \\ a & a \end{vmatrix} + d\begin{vmatrix} a & a \\ a & a \end{vmatrix} = 0$$

13. $\begin{vmatrix} 5 & -2 & 0 \\ -1 & 1 & 2 \\ 0 & -7 & 2 \end{vmatrix} = 76$

15. (a) $\det(A) = -2 = \det(A^T)$

(b) $\det(AB) = 4 = -2(-2) = \det(A)\det(B)$

(c) $\det(A^{-1}) = -\frac{1}{2} = \frac{1}{\det(A)}$

17. (a)
$M_{11} = -30 \qquad C_{11} = -30$
$M_{12} = 10 \qquad C_{12} = -10$
$M_{13} = -35 \qquad C_{13} = -35$
$M_{21} = 12 \qquad C_{21} = -12$
$M_{22} = -20 \qquad C_{22} = -20$
$M_{23} = 14 \qquad C_{23} = -14$
$M_{31} = -2 \qquad C_{31} = -2$
$M_{32} = -10 \qquad C_{32} = 10$
$M_{33} = 11 \qquad C_{33} = 11$

(b) (b_1) $1(-30) - 2(-10) + 2(-35) = -80$
(b_2) $3(-12) + 5(-20) - 4(-14) = -80$
(b_3) $7(-2) + 0(10) - 6(11) = -80$
(b_4) $1(-30) + 3(-12) + 7(-2) = -80$
(b_5) $-2(-10) + 5(-20) + 0(10) = -80$
(b_6) $2(-35) - 4(-14) - 6(11) = -80$

19. We have

$$\det(A^{-1}) = \det\begin{bmatrix} 0 & 0 & 1 & 0 \\ 0 & 1 & 0 & 0 \\ \frac{1}{2} & 0 & 0 & 0 \\ 0 & 0 & 0 & -1 \end{bmatrix} = \frac{1}{2}\begin{vmatrix} 0 & 1 & 0 \\ 1 & 0 & 0 \\ 0 & 0 & -1 \end{vmatrix} = \frac{1}{2}$$

Therefore, $\det(A^{-1}) = \dfrac{1}{2} = \dfrac{1}{\det(A)}$.

21. We have

$$\det(A^{-1}) = \begin{vmatrix} \dfrac{d}{ad-bc} & -\dfrac{b}{ad-bc} \\ -\dfrac{c}{ad-bc} & \dfrac{a}{ad-bc} \end{vmatrix} = \frac{ad-bc}{(ad-bc)^2} = \frac{1}{ad-bc} = \frac{1}{\det(A)}$$

23. $\det(E_3) = r = \det(E_4)$

25. The determinant of an elementary matrix E is $1, r$, or -1 according to whether E comes from elimination, scaling (of factor r), or interchange.

27. (a) $\det(E_3A) = r(aei - afh - bdi + cdh + bfg - ceg) = \det(E_3)\det(A) = \det(A_3)$

(b) $\det(E_4A) = r(aei - afh - bdi + cdh + bfg - ceg) = \det(E_4)\det(A) = \det(A_4)$

29. (a) $\lambda^2 - 4\lambda + 1 = 0 \Rightarrow \lambda = 2 \pm \sqrt{3}$ (b) $\lambda^2 - 11\lambda = 0 \Rightarrow \lambda = 0, 11$

31. $\begin{vmatrix} \lambda - 2 & 1 \\ 2 & \lambda - 1 \end{vmatrix} = \lambda^2 - 3\lambda = 0 \Rightarrow \lambda = 0, 3$

33. $x = x^2 - 2x + 2 \Rightarrow x = 1, 2$

35. We get the system $2a - b = 0$, $a^2 - 4 = 0$ with solutions $a = 2, b = 4$ and $a = -2, b = -4$.

37. Suppose first that T is invertible. Then the images of $\mathbf{e}_1, \mathbf{e}_2$, and $\mathbf{e}_3$ are the vectors (a, d, g), (b, e, h), and (c, f, i), respectively and they are linearly independent. So, they define a parallelepiped. By Section 2.6, this parallelepiped has volume the absolute value of

$$\det \begin{bmatrix} a & d & g \\ b & e & h \\ c & f & i \end{bmatrix} = aei - afh - bdi + bfg + cdh - ceg$$

which happens to be the same as the absolute value of the determinant of A. Hence, the volume of the image of R equals $|\det(A)|$.
If, on the other hand, T is not invertible, then the unit cube does not map onto a solid. Because, the images of the standard basis vectors in this case should map to linearly dependent vectors. Hence, the volume of the image is 0. We shall prove now that in this case the determinant of A is also 0. Because the images are linearly dependent, let us assume that (c, f, i) is a linear combination in (a, d, g) and (b, e, h), say, $(c, f, i) = k_1(a, d, g) + k_2(b, e, h)$. Then

$$\det(A) = \begin{vmatrix} a & b & k_1a + k_2b \\ d & e & k_1d + k_2e \\ g & h & k_1g + k_2h \end{vmatrix} = 0$$

Therefore, in all cases the volume of the image of the unit cube is equal to $|\det(A)|$.

Exercises 6.2

1. (a) -50 (product of diagonal entries) (b) 0 (column 1 equals column 2)

3. (a) -1 (switching rows 1 and 4) (b) -1 (switching rows 2 and 4)

5. (a) 1 (switching columns 1 and 5, then cofactor expanding about column 1)

(b) 1 (cofactor expansion about column 2)

7. 24 (switching columns 1 and 5, then taking the product of the diagonal entries)

9. Property 3 of Theorem 1 was used.

11. Property 4 of Theorem 4 was used.

13. Property 4 of Theorem 1 was used to modify the second row (did not change the determinant) and then property 2 to scale the last row.

15. Theorem 7 was used.

17. $x = 0$ makes the second row zero and $x = 2$ makes the first and last rows proportional. So, in each case the determinant should be zero.

19. (a) We have

$$\begin{vmatrix} 2 & 2 & 4 \\ -1 & -1 & -1 \\ 0 & 2 & 0 \end{vmatrix} = \begin{vmatrix} 2 & 2 & 4 \\ 0 & 0 & 1 \\ 0 & 2 & 0 \end{vmatrix} = - \begin{vmatrix} 2 & 2 & 4 \\ 0 & 2 & 0 \\ 0 & 0 & 1 \end{vmatrix} = -4$$

(b) and

$$\begin{vmatrix} 1 & 1 & 0 \\ 1 & 1 & 1 \\ 0 & 1 & 1 \end{vmatrix} = \begin{vmatrix} 1 & 1 & 0 \\ 0 & 0 & 1 \\ 0 & 1 & 1 \end{vmatrix} = - \begin{vmatrix} 1 & 1 & 0 \\ 0 & 1 & 1 \\ 0 & 0 & 1 \end{vmatrix} = -1$$

21. We have

$$\begin{vmatrix} 2 & -4 & 2 & 8 \\ -2 & 3 & 0 & -7 \\ 0 & 1 & 5 & -1 \\ 1 & 0 & 1 & 0 \end{vmatrix} = \begin{vmatrix} 2 & -4 & 2 & 8 \\ 0 & -1 & 2 & 1 \\ 0 & 1 & 5 & -1 \\ 0 & 2 & 0 & -4 \end{vmatrix} =$$

$$\begin{vmatrix} 2 & -4 & 2 & 8 \\ 0 & -1 & 2 & 1 \\ 0 & 0 & 7 & 0 \\ 0 & 0 & 4 & -2 \end{vmatrix} = \begin{vmatrix} 2 & -4 & 2 & 8 \\ 0 & -1 & 2 & 1 \\ 0 & 0 & 7 & 0 \\ 0 & 0 & 0 & -2 \end{vmatrix} = 28$$

23. We have

$$\begin{vmatrix} 1 & 0 & -1 & 2 & 1 \\ 2 & 2 & 4 & 2 & 8 \\ 0 & 0 & 3 & 0 & 3 \\ 0 & 0 & 1 & 5 & 0 \\ 0 & 1 & 0 & 1 & 0 \end{vmatrix} = \begin{vmatrix} 1 & 0 & -1 & 2 & 1 \\ 0 & 2 & 6 & -2 & 6 \\ 0 & 0 & 3 & 0 & 3 \\ 0 & 0 & 1 & 5 & 0 \\ 0 & 1 & 0 & 1 & 0 \end{vmatrix} = \begin{vmatrix} 1 & 0 & -1 & 2 & 1 \\ 0 & 2 & 6 & -2 & 6 \\ 0 & 0 & 3 & 0 & 3 \\ 0 & 0 & 1 & 5 & 0 \\ 0 & 0 & -3 & 2 & -3 \end{vmatrix} =$$

$$\begin{vmatrix} 1 & 0 & -1 & 2 & 1 \\ 0 & 2 & 6 & -2 & 6 \\ 0 & 0 & 3 & 0 & 3 \\ 0 & 0 & 0 & 5 & -1 \\ 0 & 0 & 0 & 2 & 0 \end{vmatrix} \begin{vmatrix} 1 & 0 & -1 & 2 & 1 \\ 0 & 2 & 6 & -2 & 6 \\ 0 & 0 & 3 & 0 & 3 \\ 0 & 0 & 1 & 5 & 0 \\ 0 & 0 & -3 & 2 & -3 \end{vmatrix} = \begin{vmatrix} 1 & 0 & -1 & 2 & 1 \\ 0 & 2 & 6 & -2 & 6 \\ 0 & 0 & 3 & 0 & 3 \\ 0 & 0 & 0 & 5 & -1 \\ 0 & 0 & 0 & 0 & \frac{2}{5} \end{vmatrix} = 12$$

25. (a) Expanding each side yields

$$ka_1b_2c_3 - ka_1b_3c_2 - ka_2b_1c_3 + ka_3b_1c_2 + ka_2b_3c_1 - ka_3b_2c_1$$

(b) Expanding each side yields

$$a_1b_2c_3 - a_1b_3c_2 - a_2b_1c_3 + a_3b_1c_2 + a_2b_3c_1 - a_3b_2c_1$$

27. Let A be a square matrix with ith row of the form $[c_{i1} + d_{i1} \cdots c_{in} + d_{in}]$. Then cofactor expansion of $\det(A)$ about this row yields

$$\begin{aligned} \det(A) &= (c_{i1} + d_{i1})\,C_{i1} + \cdots (c_{in} + d_{in})\,C_{in} \\ &= (c_{i1}C_{i1} + \cdots c_{in}C_{in}) + (d_{i1}C_{i1} + \cdots + d_{in}C_{in}) \end{aligned}$$

which is the sum of the determinants of the two matrices obtained from A by replacing the ith row with $[c_{i1} \cdots c_{in}]$ and with $[d_{i1} \cdots d_{in}]$. This concludes the proof of the theorem.

29. $\det(C) = -1$, so C is invertible. $\det(D) = -1$, so D is invertible.

31. $\det(G) = 3$ so G is invertible. $\det(H) = 2$ so H is invertible.

33. The matrix is noninvertible if the determinant $k^5 - 4k^3$ is zero, i.e., if $k = -2, 0, 2$.

35. We have

$$\det(B^{-1}AB) = \det(B^{-1})\det(A)\det(B) = \det(B)^{-1}\det(A)\det(B) = \det(A)$$

37. (a) -14 (b) -56 (c) 7 (d) -2 (e) -14 (f) -1134

39. We have

$$\begin{vmatrix} 1 & 1 & 1 \\ a & b & c \\ a^2 & b^2 & c^2 \end{vmatrix} = \begin{vmatrix} 1 & 1 & 1 \\ 0 & b-a & c-a \\ 0 & b^2-a^2 & c^2-a^2 \end{vmatrix} = \begin{vmatrix} 1 & 1 & 1 \\ 0 & b-a & c-a \\ 0 & 0 & (c-a)(c-b) \end{vmatrix}$$

$$= (b-a)(c-b)(c-a)$$

41. We have

$$\begin{vmatrix} 1 & 1 & 1 \\ a & b & c \\ a^3 & b^3 & c^3 \end{vmatrix} = \begin{vmatrix} 1 & 1 & 1 \\ 0 & b-a & c-a \\ 0 & b^3-a^3 & c^3-a^3 \end{vmatrix} = \begin{vmatrix} 1 & 1 & 1 \\ 0 & b-a & c-a \\ 0 & 0 & (c-b)(c-a)(a+c+b) \end{vmatrix}$$

$$= (b-a)(c-b)(c-a)(a+b+c)$$

43. If E is obtained from I by $R_i + cR_j \to R_i$, then $\det(E) = \det(I) = 1$, by Theorem 1, part 4. If E is obtained from I by $cR_i \to R_i$, then $\det(E) = \det(I) = c$, by Theorem 1, part 2. If E is obtained from I by $R_i \leftrightarrow R_j$, then $\det(E) = \det(I) = -1$, by Theorem 1, part 3.

45. If A is noninvertible so is AB, by Theorem 16, Section 3.3. But then $\det(A) = 0$ and $\det(AB) = 0$, by Theorem 3. Therefore,

$$\det(AB) = 0 = 0\det(B) = \det(A)\det(B)$$

Exercises 6.3

1. (a) $\frac{-1}{2}\begin{bmatrix} 5 & -4 \\ -3 & 2 \end{bmatrix} = \begin{bmatrix} -\frac{5}{2} & 2 \\ \frac{3}{2} & -1 \end{bmatrix}$ (b) $\frac{1}{2}\begin{bmatrix} 4 & -2 \\ 3 & -1 \end{bmatrix} = \begin{bmatrix} 2 & -1 \\ \frac{3}{2} & -\frac{1}{2} \end{bmatrix}$

3. $\frac{1}{4}\begin{bmatrix} 0 & 2 & 2 \\ 2 & -2 & 0 \\ 2 & 0 & -2 \end{bmatrix} = \begin{bmatrix} 0 & \frac{1}{2} & \frac{1}{2} \\ \frac{1}{2} & -\frac{1}{2} & 0 \\ \frac{1}{2} & 0 & -\frac{1}{2} \end{bmatrix}$

5. $\frac{1}{1}\begin{bmatrix} 1 & 0 & 0 \\ -2 & 1 & 0 \\ 5 & -4 & 1 \end{bmatrix} = \begin{bmatrix} 1 & 0 & 0 \\ -2 & 1 & 0 \\ 5 & -4 & 1 \end{bmatrix}$

7. Both row reduction

$$\begin{bmatrix} 1 & 2 & 1 & 0 & 1 & 0 & 0 & 0 \\ 2 & 2 & 1 & 1 & 0 & 1 & 0 & 0 \\ 2 & 1 & 1 & 2 & 0 & 0 & 1 & 0 \\ 1 & 2 & 1 & 1 & 0 & 0 & 0 & 1 \end{bmatrix} \sim \cdots \sim \begin{bmatrix} 1 & 0 & 0 & 0 & 0 & 1 & 0 & -1 \\ 0 & 1 & 0 & 0 & -1 & 1 & -1 & 1 \\ 0 & 0 & 1 & 0 & 3 & -3 & 2 & -1 \\ 0 & 0 & 0 & 1 & -1 & 0 & 0 & 1 \end{bmatrix}$$

and using the adjoint

$$A^{-1} = \frac{1}{-1}\begin{bmatrix} 0 & -1 & 0 & 1 \\ 1 & -1 & 1 & -1 \\ -3 & 3 & -2 & 1 \\ 1 & 0 & 0 & -1 \end{bmatrix} = \begin{bmatrix} 0 & 1 & 0 & -1 \\ -1 & 1 & -1 & 1 \\ 3 & -3 & 2 & -1 \\ -1 & 0 & 0 & 1 \end{bmatrix}$$

yield the same answer. However, the computation of the adjoint is very tedious. Row reduction is much more efficient.

9. (a) We have

$$x = \frac{\begin{vmatrix} 1 & 1 \\ 1 & -1 \end{vmatrix}}{\begin{vmatrix} 1 & 1 \\ 1 & -1 \end{vmatrix}} = 1, \quad y = \frac{\begin{vmatrix} 1 & 1 \\ 1 & 1 \end{vmatrix}}{\begin{vmatrix} 1 & 1 \\ 1 & -1 \end{vmatrix}} = \frac{0}{-2} = 0$$

(b) Also,

$$x = \frac{\begin{vmatrix} 1 & 2 \\ 1 & 4 \end{vmatrix}}{\begin{vmatrix} 1 & 2 \\ 3 & 4 \end{vmatrix}} = \frac{2}{-2} = -1, \quad z = \frac{\begin{vmatrix} 1 & 1 \\ 3 & 1 \end{vmatrix}}{\begin{vmatrix} 1 & 2 \\ 3 & 4 \end{vmatrix}} = \frac{-2}{-2} = 1$$

11. We have

$$x = \frac{\begin{vmatrix} 1 & 2 & 3 \\ 1 & 1 & 4 \\ 1 & 0 & 1 \end{vmatrix}}{\begin{vmatrix} 1 & 2 & 3 \\ 0 & 1 & 4 \\ 0 & 0 & 1 \end{vmatrix}} = \frac{4}{1} = 4, \quad y = \frac{\begin{vmatrix} 1 & 1 & 3 \\ 0 & 1 & 4 \\ 0 & 1 & 1 \end{vmatrix}}{\begin{vmatrix} 1 & 2 & 3 \\ 0 & 1 & 4 \\ 0 & 0 & 1 \end{vmatrix}} = \frac{-3}{1} = -3$$

$$z = \frac{\begin{vmatrix} 1 & 2 & 1 \\ 0 & 1 & 1 \\ 0 & 0 & 1 \end{vmatrix}}{\begin{vmatrix} 1 & 2 & 3 \\ 0 & 1 & 4 \\ 0 & 0 & 1 \end{vmatrix}} = \frac{1}{1} = 1$$

13. $z = \dfrac{-21}{-101} = \dfrac{21}{101}$

15. First we note that if A is the zero matrix, the equation is clearly true. So, we assume that A is nonzero. If $\det(A) \neq 0$, then $A\text{Adj}(A) = \det(A)I_n$ implies that

$$\begin{aligned} \det(A\text{Adj}(A)) = \det(\det(A)I_n) &\Rightarrow \det(A)\det(\text{Adj}(A)) = (\det(A))^n \\ &\Rightarrow \det(\text{Adj}(A)) = (\det(A))^{n-1} \end{aligned}$$

If $\det(A) = 0$, then $A\text{Adj}(A) = \det(A)I_n = \mathbf{0}$. But if $\text{Adj}(A)$ is invertible, then $A\text{Adj}(A) = \mathbf{0}$ would imply that $A = \mathbf{0}$, which contradicts our assumption on A. Hence, $\text{Adj}(A)$ has to be noninvertible. So, $\det(\text{Adj}(A)) = 0$. Therefore, even when $\det(A) = 0$, the identity

$$\det(\text{Adj}(A)) = (\det(A))^{n-1}$$

still holds.

17. If A is invertible, then $\det(A) \neq 0$. Hence, $\det(\mathrm{Adj}(A)) = (\det(A))^{n-1} \neq 0$, by Exercise 15. Therefore, $\mathrm{Adj}(A)$ is invertible.

Conversely, let $\mathrm{Adj}(A)$ be invertible. So, $\det(\mathrm{Adj}(A)) \neq 0$. Hence, $(\det(A))^{n-1} \neq 0$, again by Exercise 15. So, $\det(A) \neq 0$. Therefore, A is invertible.

Exercises 6.4

1. $(2,1,3,4)$ is odd with sign -1. $(1,4,2,3)$ is even with sign 1. $(1,5,2,4,3)$ is even with sign 1. $(1,4,3,5,2)$ is even with sign 1.

3. $(3,1,4,2)$ is odd with sign -1. $(4,2,1,3)$ is even with sign 1. $(3,4,2,1,5)$ is odd with sign -1. $(4,2,5,1,3)$ is even with sign 1.

5. (a) $2(3)(4) = -24$ (b) $2(3)(4) = 24$

7. (a) $2(3)(4) - 5(3)(6) = -66$ (b) $-2(3)(-4) - 5(3)(-6) = 114$

9. $1(2)(3)(4)(5) = 120$

11. We have

$$\begin{bmatrix} 1 & 0 & 0 \\ 0 & 1 & 0 \\ 0 & 0 & 1 \end{bmatrix} \leftrightarrow (1,2,3) \quad \begin{bmatrix} 1 & 0 & 0 \\ 0 & 0 & 1 \\ 0 & 1 & 0 \end{bmatrix} \leftrightarrow (1,3,2) \quad \begin{bmatrix} 0 & 1 & 0 \\ 0 & 0 & 1 \\ 1 & 0 & 0 \end{bmatrix} \leftrightarrow (3,1,2)$$

$$\begin{bmatrix} 0 & 0 & 1 \\ 0 & 1 & 0 \\ 1 & 0 & 0 \end{bmatrix} \leftrightarrow (3,2,1) \quad \begin{bmatrix} 0 & 1 & 0 \\ 1 & 0 & 0 \\ 0 & 0 & 1 \end{bmatrix} \leftrightarrow (2,1,3) \quad \begin{bmatrix} 0 & 0 & 1 \\ 1 & 0 & 0 \\ 0 & 1 & 0 \end{bmatrix} \leftrightarrow (2,3,1)$$

13. Let $p = (j_1, \ldots, j_n)$ be a permutation and let A be the corresponding permutation matrix. The ith row of A has a 1 on its j_ith column and 0s elsewhere. Hence, the determinant of A consists of only one term of the form $\mathrm{sign}(p)1 \cdots 1 = \mathrm{sign}(p)$.

15. If $kR_i \to R_i$ yields B from A, then

$$\begin{aligned}\det(B) &= \sum \pm a_{1j_1} \ldots (ka_{ij_i}) \ldots a_{nj_n} \\ &= k\sum \pm a_{1j_1} \ldots a_{ij_i} \ldots a_{nj_n} \\ &= k \det(A)\end{aligned}$$

17. We can interchange any two entries j_i, j_k $(i < k)$ in a permutation in the following fashion: We move j_i to the right by interchanging consecutive entries until is placed next to and on the left of j_k. This requires $k-i-1$ consecutive interchanges. Now we move j_k to the left in the same fashion until it reaches the ith position. This requires $k-i$ consecutive interchanges. So, we have a total of $2(k-i)-1$, i.e., an odd number of consecutive interchanges. But after each interchange the number of inversions increases or decreases by 1, by Exercise 16. Therefore, the number of inversions changes by an odd number.

19. If $R_i \leftrightarrow R_l$ $(i < l)$ yields B from A, then

$$\begin{aligned}\det(B) &= \sum \pm a_{1j_1} \ldots a_{lj_i} \ldots a_{ij_l} \ldots a_{nj_n} \\ &= \sum \pm a_{1j_1} \ldots a_{ij_l} \ldots a_{lj_i} \ldots a_{nj_n} \\ &= -\sum \pm a_{1j_1} \ldots a_{ij_i} \ldots a_{lj_l} \ldots a_{nj_n} \\ &= -\det(A)\end{aligned}$$

The third equality holds because to each permutation $(j_1, \ldots, j_l, \ldots, j_i, \ldots j_n)$ corresponds the permutation $(j_1, \ldots, j_i, \ldots, j_l, \ldots j_n)$ obtained by interchanging entries j_l and j_i and the signs of the two are opposite, by Exercise 18.

21. If $R_i + kR_l \to R_i$ yields B from A, then

$$\begin{aligned}\det(B) &= \sum \pm a_{1j_1} \ldots (a_{ij_i} + ka_{lj_i}) \ldots a_{nj_n} \\ &= \sum \pm a_{1j_1} \ldots a_{ij_i} \ldots a_{nj_n} + k\sum \pm a_{1j_1} \ldots a_{lj_i} \ldots a_{nj_n} \\ &= \sum \pm a_{1j_1} \ldots a_{ij_i} \ldots a_{nj_n} + 0 \\ &= \det(A)\end{aligned}$$

The second sum is zero, by Exercise 20, as the determinant of a matrix with equal rows (the ith and the lth).

23. Part 1 of the theorem was proved in Exercise 22. Part 2 of the theorem follows from Exercise 15 and part 1 (for columns). Part 3 of the theorem follows from

Exercise 19 and part 1 (for columns). Part 4 of the theorem follows from Exercise 21 and part 1 (for columns).

25. Let A' be the matrix obtained from A by $i-1$ successive interchanges of adjacent rows and $j-1$ successive interchanges of adjacent columns that may bring a_{ij} into the top left position maintaining the relative orders of the other elements. Then

$$\det(A) = (-1)^{i-1+j-1}\det(A') = (-1)^{i+j}\det(A')$$

by several applications of part 3 of Theorem 1. Also

$$\det(A(i,j)) = \det(A'(1,1))$$

and because $D'_{11} = C'_{11} = \det(A(i,j)) = (-1)^{ij}C_{ij}$, by Exercise 25, the sum of terms in $\det(A')$ containing $a'_{11} = a_{ij}$ is $(-1)^{ij}a_{ij}C_{ij}$. But because $\det(A') = (-1)^{i+j}\det(A)$, the sum of the terms of $\det(A')$ containing a_{ij} is also $(-1)^{i+j}a_{ij}D_{ij}$. It follows that $D_{ij} = C_{ij}$.

27. First we have $\det(A) = \det(A^T)$, by Exercise 22. Now applying the valid, by Exercise 26 cofactor expansion of $\det(A^T)$ about its jth row, yields

$$\begin{aligned}\det(A) &= \det(A^T)\\ &= a_{1j}C'_{j1} + a_{2j}C'_{j2} + \cdots + a_{nj}C'_{jn}\\ &= a_{1j}C_{1j} + a_{2j}C_{2j} + \cdots + a_{nj}C_{nj}\end{aligned}$$

because the cofactor C'_{ji} of A^T equals the cofactor C_{ij} of A. (The matrix that corresponds to C'_{ji} is the transpose of the matrix that corresponds to C_{ij}.) Hence,

$$\det(A) = a_{1j}C_{1j} + a_{2j}C_{2j} + \cdots + a_{nj}C_{nj}$$

as claimed.

Exercises 6.5

1. $\begin{vmatrix} x & y & 1 \\ -1 & 2 & 1 \\ 1 & 1 & 1 \end{vmatrix} = x + 2y - 3 = 0$

3. $\begin{vmatrix} x & y & 1 \\ 0 & 1 & 1 \\ 6 & -8 & 1 \end{vmatrix} = 9x + 6y - 6 = 0 \Rightarrow 3x + 2y - 2 = 0$

5. They are not, because $\begin{vmatrix} -2 & 0 & 1 \\ 0 & 1 & 1 \\ 2 & 1 & 1 \end{vmatrix} = -2 \neq 0.$

7. We have

$$\begin{vmatrix} x^2+y^2 & x & y & 1 \\ 0 & 0 & 0 & 1 \\ 2 & -1 & -1 & 1 \\ 4 & 0 & -2 & 1 \end{vmatrix} = 2x^2 + 2y^2 + 4y = 0$$

So, $x^2 + y^2 + 2y = 0$, or $x^2 + (y+1)^2 = 1$. The center is at $(0, -1)$ and the radius is 1.

9. We have

$$\begin{vmatrix} x^2+y^2 & x & y & 1 \\ 50 & 5 & 5 & 1 \\ 2 & 1 & -1 & 1 \\ 0 & 0 & 0 & 1 \end{vmatrix} = -10x^2 - 10y^2 + 60x + 40y$$

Hence, $x^2 + y^2 - 6x - 4y = 0$ or $(x-3)^2 + (y-2)^2 = 13$. The center is at $(3, 2)$ and the radius is $\sqrt{13}$.

11. We have

$$\begin{vmatrix} y & x^2 & x & 1 \\ 4 & 0 & 0 & 1 \\ 3 & 1 & 1 & 1 \\ 9 & 1 & -1 & 1 \end{vmatrix} = -2y + 4x^2 - 6x + 8$$

So, $y - 2x^2 + 3x - 4 = 0$, or $y = 2x^2 - 3x + 4$.

13. We have

$$\begin{vmatrix} y & x^2 & x & 1 \\ 2 & 4 & 2 & 1 \\ 1 & 9 & 3 & 1 \\ 7 & 1 & 1 & 1 \end{vmatrix} = -2y + 4x^2 - 22x + 32$$

Thus, $-2y + 4x^2 - 22x + 32 = 0$, or $y = 2x^2 - 11x + 16$.

15. Because

$$\begin{vmatrix} x & y & z & 1 \\ -1 & 1 & 1 & 1 \\ -1 & 4 & 3 & 1 \\ 4 & 0 & 2 & 1 \end{vmatrix} = 5x + 10y - 15z + 10$$

we get $x + 2y - 3z + 2 = 0$.

17. First we rewrite the system in unknown y and parametric coefficients in x.

$$\begin{aligned} y^2 + (x^2 - 1) &= 0 \\ y^2 + (x^2 - 2x - 1) &= 0 \end{aligned}$$

The Sylvester resultant is

$$\begin{vmatrix} 1 & 0 & x^2 - 1 & 0 \\ 0 & 1 & 0 & x^2 - 1 \\ 1 & -2 & x^2 + 2x + 1 & 0 \\ 0 & 1 & -2 & x^2 + 2x + 1 \end{vmatrix} = 0$$

Hence, $8x^2 + 8x = 0$. So, $x = 0$ or $x = -1$. If $x = 0$, then $y = 1$, by the second equation, and the first equation is also satisfied. If $x = -1$, then $y = 0$, by the first equation, and the second equation is also satisfied. So we have two solutions: $x = 0$, $y = 1$ and $x = -1, y = 0$.

19. From

$$\begin{vmatrix} x^2 + y^2 + z^2 & x & y & z & 1 \\ 54 & 1 & 2 & 7 & 1 \\ 38 & 5 & 2 & 3 & 1 \\ 46 & 1 & 6 & 3 & 1 \\ 6 & 1 & 2 & -1 & 1 \end{vmatrix} = 0$$

we get after simplification

$$x^2 + y^2 + z^2 - 2x - 4y - 6z - 2 = 0$$

and completion of squares

$$(x-1)^2 + (y-2)^2 + (z-3)^2 = 16$$

we see that the sphere has center $(1, 2, 3)$ and radius 4.

21. The underlying graph is

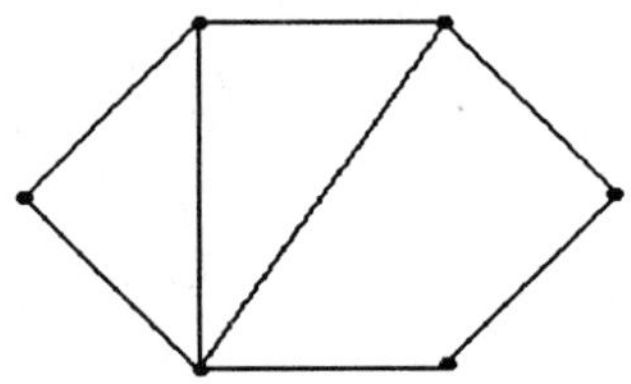

23. (a) We have the following three spanning trees

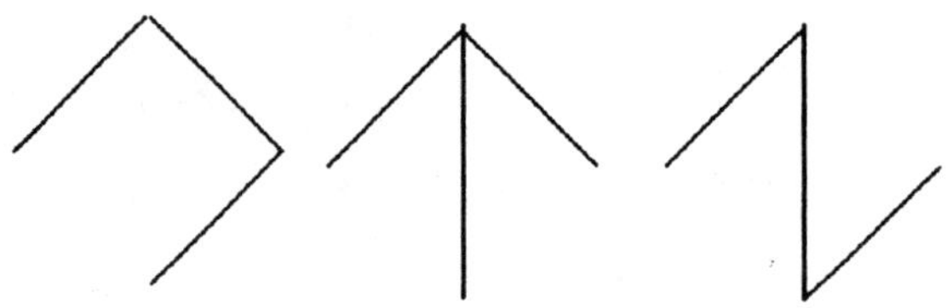

(b) The tree matrix is

$$\begin{bmatrix} 3 & -1 & -1 & -1 \\ -1 & 1 & 0 & 0 \\ 0 & -1 & 2 & -1 \\ -1 & 0 & -1 & 2 \end{bmatrix}$$

and the cofactor

$$C_{11} = (-1)^2 \begin{vmatrix} 1 & 0 & 0 \\ -1 & 2 & -1 \\ 0 & -1 & 2 \end{vmatrix} = 3$$

Hence, there are three spanning trees.

Exercises 7.1

1. $A\mathbf{u} = (-5, 5) = 5\mathbf{u}$. The eigenvalue is 5.

3. $A\mathbf{v} = (0, 0) = 0\mathbf{v}$. The eigenvalue is 0.

5. No, because $A(\mathbf{u} + \mathbf{v}) = A(1, 4) = (-5, 5)$ which is not a multiple of $(1, 4)$.

7. Vector $(-1, 1)$ is not an eigenvector of the reduced row echelon form $\begin{bmatrix} 1 & -\frac{2}{3} \\ 0 & 0 \end{bmatrix}$ of A. On the other hand, vector $(2, 3)$ is an eigenvector with eigenvalue 0.

9. Only $(-2, 1)$ and $(1, -\frac{1}{2})$ are eigenvectors. The corresponding eigenvalue for both is 0.

11. There are two ways of showing that $-6, 4, 7$ are eigenvalues of A. First, using the definition, it suffices to show that each of the homogeneous systems

$$(A - (-6)\,I)\,\mathbf{x} = \mathbf{0}, \qquad (A - 4I)\,\mathbf{x} = \mathbf{0}, \qquad (A - 7I)\,\mathbf{x} = \mathbf{0}$$

has nontrivial solutions. For example, the first system

$$\begin{bmatrix} 7+6 & 0 & 0 \\ 0 & -2+6 & 4 \\ 0 & 6 & 0+6 \end{bmatrix} \begin{bmatrix} x \\ y \\ z \end{bmatrix} = \begin{bmatrix} 0 \\ 0 \\ 0 \end{bmatrix}$$

yields $13x = 0$, $4y + 4z = 0$, $6y + 6z = 0$, which does have nontrivial solutions. Likewise, the other two systems have nontrivial solutions.

The second way is to simply compute and solve the characteristic equation of A.

$$\begin{vmatrix} 7-\lambda & 0 & 0 \\ 0 & -2-\lambda & 4 \\ 0 & 6 & 0-\lambda \end{vmatrix} = (7-\lambda)\left(\lambda^2 + 2\lambda - 24\right) = 0$$

The solutions are the eigenvalues, $-6, 4, 7$.

13. The only vectors that remain on their own line after this projection are the ones along the two axes. Along the x–axis the vectors remain the same, so their eigenvalue is 1. Along the y–axis the vectors go to the origin, so their eigenvalue is 0.

15. This transformation has no eigenvectors (with real entries), because all nonzero vectors get rotated, so none remains in its own line.

17. We have:

(a) The characteristic polynomial is $\lambda^2 + 2\lambda - 24$. The eigenvalues are 4 and -6. The corresponding bases of eigenvectors are $\{(2,3)\}$ and $\{(1,-1)\}$. All multiplicities are 1.

(b) The characteristic polynomial is $\lambda^2 + 6\lambda + 9$. The only eigenvalue is -3. The corresponding basis of eigenvectors is $\{(3,1)\}$. The algebraic multiplicity is 2. The geometric multiplicity is 1.

(c) The characteristic polynomial is $\lambda^2 - 2\lambda - 35$. The eigenvalues are 7 and -5. The corresponding bases of eigenvectors are $\{(1,1)\}$ and $\{(-7,5)\}$. All multiplicities are 1.

19. We have:

(a) The characteristic polynomial is $-\lambda^3 + \lambda^2 + \lambda - 1$. The eigenvalues are 1 with basis of eigenvectors $\{(1,0,1), (0,1,0)\}$, and -1 with basis of eigenvectors $\{(-1,0,1)\}$. The algebraic and geometric multiplicity of 1 is 2. The algebraic and geometric multiplicity of -1 is 1.

(b) The characteristic polynomial is $(1-\lambda)(2-\lambda)(3-\lambda)$. The eigenvalues are $1, 2, 3$ with corresponding eigenvector bases $\{(1,0,0)\}, \{(1,1,0)\}, \{(0,0,1)\}$. All multiplicities are 1.

21. We have:

(a) The characteristic polynomial is $-\lambda^3 + 3\lambda^2 + 4\lambda - 12$. The eigenvalues are $2, 3$ and -2 with corresponding bases of eigenvectors $\{(1,1,0)\}$, $\{(0,0,1)\}$ and $\{(-1,1,0)\}$. All multiplicities are 1.

(b) The characteristic polynomial is $(1-\lambda)(\lambda^2 - 1)$. The eigenvalues are 1 with basis of eigenvectors $\{(1,0,0), (0,1,1)\}$, and -1 with basis of eigenvectors $\{(0,-1,1)\}$. The algebraic and geometric multiplicity of 1 is 2. The algebraic and geometric multiplicity of -1 is 1.

23. We have:

(a) The characteristic polynomial is $(1-\lambda)(2-\lambda)(3-\lambda)$. The eigenvalues are $1, 2, 3$ with corresponding eigenvector bases $\{(2,-3,3)\}, \{(0,1,0)\}, \{(0,1,1)\}$. All multiplicities are 1.

(b) The characteristic polynomial is $(1-\lambda)(2-\lambda)(3-\lambda)$. The eigenvalues are $1, 2, 3$ with corresponding eigenvector bases $\{(1,-2,2)\}$, $\{(0,1,0)\}$, $\{(0,1,1)\}$. All multiplicities are 1.

25. $\begin{bmatrix} 1 & 1 & 1 \\ 0 & 2 & 1 \\ 0 & 0 & 1 \end{bmatrix}$ with eigenvalues 1 and 2.

27. (a) $1, -2$ (b) $-4, 4$

29. $(1,1,1)$ with eigenvalue $a+b+c$.

31. For any square matrix B we have $\det(B) = \det(B^T)$. Therefore,

$$\det(A - \lambda I) = \det((A-\lambda I)^T) = \det(A^T - \lambda I^T) = \det(A^T - \lambda I)$$

So, A and A^T have the same characteristic polynomial. Thus, they have the same eigenvalues.

33. One solution is $\frac{1}{2}\mathbf{v}$ because

$$A\left(\frac{1}{2}\mathbf{v}\right) = \frac{1}{2}A\mathbf{v} = \frac{1}{2}2\mathbf{v} = \mathbf{v}$$

35. First we show that if A is invertible, then $\lambda \neq 0$. For if $\lambda = 0$, then $A\mathbf{v} = 0\mathbf{v} = \mathbf{0}$. So, $A^{-1}A\mathbf{v} = \mathbf{0}$, or $\mathbf{v} = \mathbf{0}$, which contradicts the assumption that $\mathbf{v}$ is an eigenvector. Now, $A\mathbf{v} = \lambda\mathbf{v}$ implies $A^{-1}A\mathbf{v} = A^{-1}\lambda\mathbf{v}$, or $\mathbf{v} = \lambda A^{-1}\mathbf{v}$. Hence, $A^{-1}\mathbf{v} = \lambda^{-1}\mathbf{v}$, because $\lambda \neq 0$. Therefore, $\mathbf{v}$ is an eigenvector of A^{-1} with eigenvalue λ^{-1}.

37. We have the following.

(a)

$$\begin{aligned} \det(B - \lambda I) &= \det(P^{-1}AP - \lambda I) \\ &= \det(P^{-1}AP - \lambda P^{-1}IP) \\ &= \det(P^{-1}(A - \lambda I)P) \\ &= \det(P^{-1})\det(A - \lambda I)\det(P) \\ &= \det(P)^{-1}\det(A - \lambda I)\det(P) \\ &= \det(A - \lambda I) \end{aligned}$$

(b) A and B have the same eigenvalues because they have the same characteristic polynomial by part (a).

(c) Because

$$B\mathbf{v} = \lambda\mathbf{v} \Leftrightarrow P^{-1}AP\mathbf{v} = \lambda\mathbf{v} \Leftrightarrow AP\mathbf{v} = \lambda P\mathbf{v}$$

$P\mathbf{v}$ is an eigenvector of A with eigenvalue λ. ($P\mathbf{v}$ is nonzero because $\mathbf{v}$ is nonzero and P is invertible.)

(d) Because $A\mathbf{u} = \lambda\mathbf{u}$ and $B = P^{-1}AP$, we have

$$BP^{-1}\mathbf{u} = P^{-1}APP^{-1}\mathbf{u} = P^{-1}A\mathbf{u} = P^{-1}A\mathbf{u} = P^{-1}\lambda\mathbf{u} = \lambda P^{-1}\mathbf{u}$$

Thus, $P^{-1}\mathbf{u}$ is an eigenvector of B with eigenvalue λ.

39. $A\mathbf{v} = \mathbf{0} = 0\mathbf{v}$, because $\mathbf{v} \in \text{Null}(A)$. Therefore, $\mathbf{v} \neq \mathbf{0}$ is an eigenvector with eigenvalue 0.

41. Let $A = \begin{bmatrix} 1 & 1 \\ 0 & 1 \end{bmatrix}$ and $B = \begin{bmatrix} 0 & 1 \\ 1 & 0 \end{bmatrix}$. The only eigenvalue of A is 1. The eigenvalues of B are 1 and -1. The eigenvalues of the product $AB = \begin{bmatrix} 1 & 1 \\ 1 & 0 \end{bmatrix}$ are $\frac{1}{2} + \frac{1}{2}\sqrt{5}, \frac{1}{2} - \frac{1}{2}\sqrt{5}$, neither of which is the product of the eigenvalues of A and B.

43. Because every nonzero n-vector is an eigenvector of A, $\mathbf{e}_1, \ldots, \mathbf{e}_n$ are eigenvectors. Let $\lambda_1, \ldots, \lambda_n$ be the corresponding eigenvalues. So, $A\mathbf{e}_i = \lambda_i\mathbf{e}_i$. Hence, the ith column of A is $\lambda_i\mathbf{e}_i$. Therefore, A is diagonal with diagonal entries $\lambda_1, \ldots, \lambda_n$. Next, we show that all the λ_i's are equal. Let us consider λ_i and λ_j for $i \neq j$. Because $\mathbf{e}_i + \mathbf{e}_j$ is an eigenvector with eigenvalue, say k, then

$$A(\mathbf{e}_i + \mathbf{e}_j) = k(\mathbf{e}_i + \mathbf{e}_j) = k\mathbf{e}_i + k\mathbf{e}_j$$

On the other hand,

$$A(\mathbf{e}_i + \mathbf{e}_j) = A\mathbf{e}_i + A\mathbf{e}_j = \lambda_i\mathbf{e}_i + \lambda_j\mathbf{e}_j$$

Therefore, $\lambda_i\mathbf{e}_i + \lambda_j\mathbf{e}_j = k\mathbf{e}_i + k\mathbf{e}_j$, which implies $\lambda_i = k$ and $\lambda_j = k$, because $\{\mathbf{e}_i, \mathbf{e}_j\}$ is linearly independent. So, $\lambda_i = \lambda_j$. Therefore, all the diagonal entries of A are equal. Hence, A is a scalar matrix.

45. Because A is nilpotent 0 is its only eigenvalue, by the preceding exercise. The geometric multiplicity of 0 is, by definition, the dimension of E_0. But in Exercise 32 it was shown that $\text{Null}(A) = E_0$. Hence, the geometric multiplicity of 0 equals the dimension of $\text{Null}(A)$, which is the nullity of A.

47. $C(p) = \begin{bmatrix} 0 & 1 \\ 15 & -2 \end{bmatrix}$. The characteristic polynomial is $\lambda^2 + 2\lambda - 15$.

49. By the preceding exercise it suffices to construct a monic polynomial with roots 4 and -5 and then take its companion matrix. So,

$$p(x) = (x-4)(x+5) = x^2 + x - 20$$

Hence, $C(p) = \begin{bmatrix} 0 & 1 \\ 20 & -1 \end{bmatrix}$ must have eigenvalues: $4, -5$.

51. The polynomial $(x-4)(x+5)(x+2) = x^3 + 3x^2 - 18x - 40$ has companion matrix

$$C(p) = \begin{bmatrix} 0 & 1 & 0 \\ 0 & 0 & 1 \\ 40 & 18 & -3 \end{bmatrix}$$

Its eigenvalues are the roots of its characteristic polynomial $-\lambda^3 - 3\lambda^2 + 18\lambda + 40$ which are 4, -5, and -2.

53. For $n = 2$ the companion matrix of $x^2 + a_1 x + a_2$ is $\begin{bmatrix} 0 & 1 \\ -a_1 & -a_2 \end{bmatrix}$ and has characteristic polynomial $\lambda^2 + a_1\lambda + a_2$ which validates the claim in this case. Let us assume now that this claim is true for all polynomials up to degree $n-1$. The companion matrix $C(p)$ of $p(x) = x^n + a_{n-1}x^{n-1} + \cdots + a_0$ is

$$C(p) = \begin{bmatrix} 0 & 1 & 0 & \cdots & 0 \\ 0 & 0 & 1 & \cdots & 0 \\ \vdots & \vdots & \vdots & \vdots & \vdots \\ 0 & 0 & 0 & \cdots & 1 \\ -a_0 & -a_1 & -a_2 & \cdots & -a_{n-1} \end{bmatrix}$$

So, its characteristic polynomial is the determinant

$$\begin{vmatrix} -\lambda & 1 & 0 & \cdots & 0 \\ 0 & -\lambda & 1 & \cdots & 0 \\ \vdots & \vdots & \vdots & \vdots & \vdots \\ 0 & 0 & 0 & \cdots & 1 \\ -a_0 & -a_1 & -a_2 & \cdots & -a_{n-1} - \lambda \end{vmatrix}$$

which expanded about the first column yields

$$-\lambda \begin{vmatrix} -\lambda & 1 & \cdots & 0 \\ 0 & -\lambda & \cdots & \\ \vdots & \vdots & \vdots \ \ddots & 1 \\ -a_1 & -a_2 & \cdots & -a_{n-1}-\lambda \end{vmatrix} + (-1)^n a_0 \begin{vmatrix} 1 & 0 & \cdots & 0 \\ -\lambda & 1 & \cdots & 0 \\ \vdots & \vdots & \vdots & \vdots \\ 0 & 0 & \cdots & 1 \end{vmatrix}$$

$$= -\lambda(-1)^{n-1}(\lambda^{n-1} + a_{n-1}\lambda^{n-2} + \cdots + a_1) + (-1)^n a_0 \cdot 1$$

$$= (-1)^n(\lambda^n + a_{n-1}\lambda^{n-1} + \cdots + a_0)$$

$$= (-1)^n p(\lambda)$$

because by induction the first determinant is $(-1)^{n-1}(\lambda^{n-1}+a_{n-1}\lambda^{n-2}+\cdots+a_1)$. We have proved the claim for n so it is valid for any positive integer n.

55. First we find the eigenvalues and eigenvectors of the matrix $A = \begin{bmatrix} 0 & 1 \\ 1 & 0 \end{bmatrix}$ of T with respect to the standard basis $\{1, x\}$. The eigenvalues are 1 and -1 with corresponding basic eigenvectors $(1,1)$ and $(-1,1)$. Hence, T has eigenvalues 1 and -1 with corresponding basic eigenvectors $x+1$ and $-x+1$.

Exercises 7.2

---

1. The eigenvalues are $3, -7$ with corresponding basic eigenvectors $(1,1)$ and $(-1,1)$. Therefore, the matrix is diagonalizable and can be diagonalized by $P = \begin{bmatrix} 1 & -1 \\ 1 & 1 \end{bmatrix}$ and $D = \begin{bmatrix} 3 & 0 \\ 0 & -7 \end{bmatrix}$.

3. The eigenvalues are $3, 1, -7$ with corresponding basic eigenvectors $(1,1,1)$, $(1,0,1)$ and $(1,4,-4)$. Therefore, the matrix is diagonalizable, diagonalized by

$$P = \begin{bmatrix} 1 & 1 & 1 \\ 1 & 0 & 4 \\ 1 & 0 & -4 \end{bmatrix}, \qquad D = \begin{bmatrix} 3 & 0 & 0 \\ 0 & 1 & 0 \\ 0 & 0 & -7 \end{bmatrix}$$

5. The matrix is not diagonalizable. The eigenvalues are -4 and 1 with corresponding eigenspace bases $\{(25, 2, -12)\}$ and $\{(0, 1, -1)\}$. Thus, there are no three linearly independent eigenvectors.

7. $A(10, 0) = (-10, 0)$, so $(10, 0)$ is an eigenvector with eigenvalue -1. $A(6, 5) = (24, 20)$, so $(6, 5)$ is an eigenvector with eigenvalue 4. Because $(10, 0)$ and $(6, 5)$ belong to different eigenvalues they are linearly independent. A is diagonalizable, diagonalized by $P = \begin{bmatrix} 10 & 6 \\ 0 & 5 \end{bmatrix}$ and $D = \begin{bmatrix} -1 & 0 \\ 0 & 4 \end{bmatrix}$.

9. Because

$$\begin{bmatrix} 0 & 0 & 1 \\ 0 & 2 & 0 \\ 4 & 0 & 0 \end{bmatrix} \begin{bmatrix} 0 & 1 & 1 \\ 2 & 0 & 0 \\ 0 & 2 & -2 \end{bmatrix} = \begin{bmatrix} 0 & 2 & -2 \\ 4 & 0 & 0 \\ 0 & 4 & 4 \end{bmatrix}$$

we see that $(0, 2, 0), (1, 0, 2), (1, 0, -2)$ are eigenvectors with corresponding eigenvalues $2, 2, -2$. The first two eigenvectors are clearly linearly independent. Thus, S is linearly independent because $(1, 0, -2)$ belongs to a different eigenvalue. So, A is diagonalizable, diagonalized by

$$P = \begin{bmatrix} 0 & 1 & 1 \\ 2 & 0 & 0 \\ 0 & 2 & -2 \end{bmatrix}, \qquad D = \begin{bmatrix} 2 & 0 & 0 \\ 0 & 2 & 0 \\ 0 & 0 & -2 \end{bmatrix}$$

11. Let $P = \begin{bmatrix} 1 & 1 & 0 \\ 0 & 1 & 0 \\ 0 & 0 & 1 \end{bmatrix}$, $D = \begin{bmatrix} 1 & 0 & 0 \\ 0 & 2 & 0 \\ 0 & 0 & 3 \end{bmatrix}$. Then

$$A = PDP^{-1} = \begin{bmatrix} 1 & 1 & 0 \\ 0 & 2 & 0 \\ 0 & 0 & 3 \end{bmatrix}$$

13. Let $P = \begin{bmatrix} 1 & -3 & -2 \\ 1 & 0 & 1 \\ 1 & 1 & 0 \end{bmatrix}$, $D = \begin{bmatrix} 6 & 0 & 0 \\ 0 & 0 & 0 \\ 0 & 0 & 0 \end{bmatrix}$. Then

$$A = PDP^{-1} = \begin{bmatrix} 1 & 2 & 3 \\ 1 & 2 & 3 \\ 1 & 2 & 3 \end{bmatrix}$$

15. A has three distinct eigenvalues, so it is diagonalizable, by Theorem 8.

17. $\{(1,1,1),(-2,1,0),(-2,0,1)\}$ (corresponding to eigenvalues $5,0,0$)

19. A is diagonalizable because its has three linearly independent eigenvectors, namely, $(1,1,1)$, $(-3,2,0)$, $(-2,0,1)$ (corresponding to eigenvalues $9,0,0$).

21. A is not diagonalizable because its only eigenvalue 5 has one (less than three) basic eigenvector. ($\{(1,0,0)\}$ is a basis for E_5.)

23. The characteristic polynomial of A is $-\lambda^3+(a+b+c)\lambda^2$. So, the eigenvalues are 0, or $a+b+c$. If $a+b+c=0$, then 0 is the only eigenvalue and the basic eigenvectors of

$$\begin{bmatrix} a & b & -a-b \\ a & b & -a-b \\ a & b & -a-b \end{bmatrix}$$

are $((a+b)/a,0,1)$ and $(-b/a,1,0)$, if $a\neq 0$. If $a=0$, then the only eigenvectors are $(1,0,0)$ and $(0,1,1)$ (because b cannot be zero in this case). So, if $a+b+c=0$, then A is not diagonalizable.

If, on the other hand, $a+b+c\neq 0$, then 0 and $a+b+c$ are distinct eigenvalues. $(1,1,1)$ is a basic eigenvector corresponding to $a+b+c$ and $((a+b)/a,0,1)$ and $(-b/a,1,0)$ are the basic eigenvectors corresponding to 0, if $a\neq 0$. If $a=0$, then $(1,0,0)$ and $(0,-c/b,1)$ are the basic eigenvectors corresponding to 0. Therefore, if $a+b+c\neq 0$, then A has always three linearly independent eigenvectors, so it is diagonalizable.

25. The characteristic polynomial of the matrix is $(1-\lambda)(\lambda^2-a)$. If $a<0$, then the only real eigenvalue is 1 with only basic eigenvector $(1,0,0)$, so the matrix is not diagonalizable. If $a=0$, then the eigenvalues are 0 and 1 with only basic eigenvectors $(0,1,0)$ and $(1,0,0)$. So, in this case the matrix is not diagonalizable, either. Let $a>0$. If $a=1$, then the eigenvalues are $-1,1$ with eigenvectors $(0,-1,1)$ and $(1,0,0)$, $(0,1,1)$. So, in this case the matrix is diagonalizable. Lastly, if $a>0$ and not 1, then we have three distinct eigenvalues, thus the matrix is again diagonalizable. So, we have proved that the matrix is diagonalizable over the reals if and only if $a>0$.

27. We have

$$A^6=\begin{bmatrix} 2 & -2 \\ 1 & 1 \end{bmatrix}\begin{bmatrix} 4 & 0 \\ 0 & -4 \end{bmatrix}^6\begin{bmatrix} 2 & -2 \\ 1 & 1 \end{bmatrix}^{-1}=\begin{bmatrix} 4096 & 0 \\ 0 & 4096 \end{bmatrix}$$

Note that $A^6 = \begin{bmatrix} 2^{12} & 0 \\ 0 & 2^{12} \end{bmatrix}$. Also,

$$A^9 = \begin{bmatrix} 2 & -2 \\ 1 & 1 \end{bmatrix} \begin{bmatrix} 4 & 0 \\ 0 & -4 \end{bmatrix}^9 \begin{bmatrix} 2 & -2 \\ 1 & 1 \end{bmatrix}^{-1} = \begin{bmatrix} 0 & 524288 \\ 131072 & 0 \end{bmatrix}$$

Note that $A^9 = \begin{bmatrix} 0 & 2^{19} \\ 2^{17} & 0 \end{bmatrix}$.

29. $P = \begin{bmatrix} 0 & 1 & 2 \\ 1 & 0 & 1 \\ -1 & -1 & 2 \end{bmatrix}$, $D = \begin{bmatrix} 0 & 0 & 0 \\ 0 & 0 & 0 \\ 0 & 0 & 5 \end{bmatrix}$ diagonalize the matrix. Hence,

$$PD^7P^{-1} = \begin{bmatrix} 2\cdot 5^6 & 2\cdot 5^6 & 2\cdot 5^6 \\ 5^6 & 5^6 & 5^6 \\ 2\cdot 5^6 & 2\cdot 5^6 & 2\cdot 5^6 \end{bmatrix} = \begin{bmatrix} 31250 & 31250 & 31250 \\ 15625 & 15625 & 15625 \\ 31250 & 31250 & 31250 \end{bmatrix}$$

31. The identity is true because by diagonalizing the base matrix on the left we get

$$\begin{bmatrix} 1 & 1 \\ 1 & 1 \end{bmatrix}^{k+1} = \begin{bmatrix} 1 & -1 \\ 1 & 1 \end{bmatrix} \begin{bmatrix} 2 & 0 \\ 0 & 0 \end{bmatrix}^{k+1} \begin{bmatrix} 1 & -1 \\ 1 & 1 \end{bmatrix}^{-1} = \begin{bmatrix} 2^k & 2^k \\ 2^k & 2^k \end{bmatrix}$$

for $k = 0, 1, 2, \ldots$.

33. Let $\lambda_1, \ldots, \lambda_k$ be all distinct eigenvalues of an $n \times n$ matrix A and let M_{λ_i} and m_{λ_i} be the corresponding algebraic and geometric multiplicities of λ_i.

First, let us assume that A is diagonalizable. So, A has n linearly independent eigenvectors. Therefore, the sum of the geometric multiplicities is n. But the sum of all algebraic multiplicities equals the degree of the characteristic polynomial, which is always n. Hence,

$$\sum m_{\lambda_i} = n = \sum M_{\lambda_i}$$

But for all i we have $M_{\lambda_i} \geq m_{\lambda_i}$, from Exercise 38, Section 7.1. Therefore, $M_{\lambda_i} = m_{\lambda_i}$. So, for each eigenvalue the algebraic and geometric multiplicities are equal.

Conversely, if we assume that for all $i = 1, \ldots, k$ the algebraic and geometric multiplicities are equal, so $M_{\lambda_i} = m_{\lambda_i}$, then

$$\sum m_{\lambda_i} = \sum M_{\lambda_i} = n$$

Hence, A has n linearly independent eigenvectors thus, it is diagonalizable. This completes the proof of the theorem.

35. Let $\mathcal{B} = \{1, x, x^2, x^3\}$ be the standard basis of P_3. Then the matrix of $T : P_3 \to P_3$

$$T(a + bx + cx^2 + dx^3) = a$$

with respect to $\mathcal{B}$ is

$$A = \begin{bmatrix} 1 & 0 & 0 & 0 \\ 0 & 0 & 0 & 0 \\ 0 & 0 & 0 & 0 \\ 0 & 0 & 0 & 0 \end{bmatrix}.$$

A is diagonalizable. Its eigenvalues are $0, 1$ with corresponding bases of eigenvectors $\{\mathbf{e}_2, \mathbf{e}_3, \mathbf{e}_4\}$ and $\{\mathbf{e}_1\}$. Therefore, T is diagonalizable. $\mathcal{B}$ is a basis of P_3 that consists of eigenvectors of T.

37. Let T be the given reflection. Then the standard matrix of T is $A = \begin{bmatrix} 1 & 0 \\ 0 & -1 \end{bmatrix}$, which has eigenvalues $1, -1$ and corresponding bases of eigenvectors $\{(1, 0)\}$, $\{(0, 1)\}$. Therefore, A is diagonalizable. Hence, T is diagonalizable. The standard basis $\{\mathbf{e}_1, \mathbf{e}_2\}$ is a basis of $\mathbf{R}^2$ that consists of eigenvectors of T.

39. Let T be the given projection. The standard matrix of T is $A = \begin{bmatrix} 1 & 0 \\ 0 & 0 \end{bmatrix}$,which has eigenvalues $0, 1$ and corresponding bases of eigenvectors $\{(0, 1)\}$ and $\{(1, 0)\}$. Therefore, A is diagonalizable. Hence, T is diagonalizable. The standard basis $\{\mathbf{e}_1, \mathbf{e}_2\}$ is a basis of $\mathbf{R}^2$ that consists of eigenvectors of T.

41. Let T be the given projection. The standard matrix of T is

$$A = \begin{bmatrix} 1 & 0 & 0 \\ 0 & 1 & 0 \\ 0 & 0 & 0 \end{bmatrix}$$

which has eigenvalues $0, 1$ and corresponding bases of eigenvectors $\{(0, 0, 1)\}$ and $\{(1, 0, 0), (0, 1, 0)\}$. Therefore, A is diagonalizable. Hence, T is diagonalizable. The standard basis $\{\mathbf{e}_1, \mathbf{e}_2, \mathbf{e}_3\}$ is a basis of $\mathbf{R}^3$ that consists of eigenvectors of T.

Exercises 7.3

---

1. 5 is an eigenvalue with eigenvector $(1, 1)$. Because $4687/937 = 5.0021$, we estimate the eigenvalue to be 5. The corresponding eigenvector is $(1, 1)$.

3. $A^6\mathbf{x} = \mathbf{u}$ and $A^7\mathbf{x} = \mathbf{v}$ imply that $A\mathbf{u} = \mathbf{v} = 4\mathbf{u}$. Thus, $\mathbf{u} = (2, 1)$ is an eigenvector with corresponding eigenvalue $\lambda = 4$.

5. Because $A^{-6}\mathbf{x} = \mathbf{u}$ and $A^{-7}\mathbf{x} = \mathbf{v}$, we have

$$A^{-6}\mathbf{x} = \mathbf{u} = A\left(A^{-7}\mathbf{x}\right) = A\mathbf{v} = 4A\mathbf{u}$$

So, $A\mathbf{u} = \frac{1}{4}\mathbf{u}$. Thus, $(2, 1)$ is an eigenvector with corresponding eigenvalue $\dfrac{1}{4}$.

7. For $\mathbf{x} = (1, 2)$ we have $A^3\mathbf{x} = \begin{bmatrix} 61 \\ -64 \end{bmatrix}$, $A^4\mathbf{x} = \begin{bmatrix} -311 \\ 314 \end{bmatrix}$. So,

$$\lambda_{\mathbf{appr}} = -5.0984, \quad \lambda = -5, \qquad \mathbf{v}_{\mathbf{appr}} = \begin{bmatrix} -0.99 \\ 1 \end{bmatrix}, \quad \mathbf{v} = \begin{bmatrix} -1 \\ 1 \end{bmatrix}$$

9. For $\mathbf{x} = (1, 2)$ we have $A^3\mathbf{x} = \begin{bmatrix} 170 \\ -173 \end{bmatrix}$, $A^4\mathbf{x} = \begin{bmatrix} -1199 \\ 1202 \end{bmatrix}$. So,

$$\lambda_{\mathbf{appr}} = -7.0529, \quad \lambda = -7, \qquad \mathbf{v}_{\mathbf{appr}} = \begin{bmatrix} -0.997 \\ 1 \end{bmatrix}, \quad \mathbf{v} = \begin{bmatrix} -1 \\ 1 \end{bmatrix}$$

11. For $\mathbf{x} = (1, 2)$ we have $A^3\mathbf{x} = \begin{bmatrix} 363 \\ -366 \end{bmatrix}$, $A^4\mathbf{x} = \begin{bmatrix} -3279 \\ 3282 \end{bmatrix}$. So,

$$\lambda_{\mathbf{appr}} = -9.0331, \quad \lambda = -9, \qquad \mathbf{v}_{\mathbf{appr}} = \begin{bmatrix} -0.999 \\ 1 \end{bmatrix}, \quad \mathbf{v} = \begin{bmatrix} -1 \\ 1 \end{bmatrix}$$

13. For $\mathbf{x} = (1, 2)$ we have $A^3\mathbf{x} = \begin{bmatrix} 364 \\ -148 \end{bmatrix}$, $A^4\mathbf{x} = \begin{bmatrix} -2924 \\ 1172 \end{bmatrix}$. So,

$$\lambda_{\mathbf{appr}} = -8.033, \quad \lambda = -8, \qquad \mathbf{v}_{\mathbf{appr}} = \begin{bmatrix} 1 \\ -0.4008 \end{bmatrix}, \quad \mathbf{v} = \begin{bmatrix} 1 \\ -\frac{2}{5} \end{bmatrix}$$

15. For $\mathbf{x} = (1, 2)$ we have $A^3\mathbf{x} = \begin{bmatrix} 2395 \\ 1598 \end{bmatrix}$, $A^4\mathbf{x} = \begin{bmatrix} 26353 \\ 17570 \end{bmatrix}$. So,

$$\lambda_{\text{appr}} = 11.003, \quad \lambda = 11, \qquad \mathbf{v}_{\text{appr}} = \begin{bmatrix} 1 \\ 0.66672 \end{bmatrix}, \quad \mathbf{v} = \begin{bmatrix} 1 \\ \frac{2}{3} \end{bmatrix}$$

17. For $\mathbf{x} = (1, 2)$ we have $A^3\mathbf{x} = \begin{bmatrix} 1412 \\ -316 \end{bmatrix}$, $A^4\mathbf{x} = \begin{bmatrix} -16964 \\ 3772 \end{bmatrix}$. So,

$$\lambda_{\text{appr}} = -12.014, \quad \lambda = -12, \qquad \mathbf{v}_{\text{appr}} = \begin{bmatrix} 1 \\ -0.22235 \end{bmatrix}, \quad \mathbf{v} = \begin{bmatrix} 1 \\ -\frac{2}{9} \end{bmatrix}$$

19. For $\mathbf{x} = (1, 2)$ we have $A^3\mathbf{x} = \begin{bmatrix} 1265 \\ -1479 \end{bmatrix}$, $A^4\mathbf{x} = \begin{bmatrix} -17729 \\ 20687 \end{bmatrix}$. So,

$$\lambda_{\text{appr}} = -14.015, \quad \lambda = -14, \qquad \mathbf{v}_{\text{appr}} = \begin{bmatrix} -0.85701 \\ 1 \end{bmatrix}, \quad \mathbf{v} = \begin{bmatrix} -\frac{6}{7} \\ 1 \end{bmatrix}$$

21. Starting at $(1, 2)$ the first four iterations yield -1.4000, -3.9412, -4.9432, -4.9977. So, the dominant eigenvalue is -5.

23. Starting at $(1, 2)$ the first four iterations yield -1.6000, -6.0690, -6.9776, -6.9995. So, the dominant eigenvalue is -7.

25. Starting at $(1, 2)$ the first four iterations yield 0.2800, 0.7882, 0.9886, 0.9995. The true eigenvalue closest to the origin is 1 and is approximated by 0.9995.

27. Starting at $(1, 2)$ the first four iterations yield 0.2286, 0.8670, 0.9968, 0.9999. The true eigenvalue closest to the origin is 1 and is approximated by 0.9999.

29. Starting at $(1, 2)$ the first four iterations yield 0.2000, 0.9111, 0.9988, 1.0000. The true eigenvalue closest to the origin is 1 and is approximated to four decimal places by 1.0000.

31. Starting at $(1, 2)$ the first four iterations yield -1.1111, -1.0194, -1.0022, -1.0022. The true eigenvalue closest to the origin is -1 and is approximated by -1.0022.

33. Starting at $(1, 2)$ the first four iterations yield -1.1000, -1.0205, -1.0026, -1.0003. The true eigenvalue closest to the origin is -1 and is approximated by -1.0003.

35. Applying the inverse power method to $\begin{bmatrix} 0 & 1 \\ -4 & 5 \end{bmatrix}$ starting at $(1,0)$, yields -1.0139 after 4 iterations. So, the root nearest 5 is $5 + (-1.0139)^{-1} = 4.0138$.

37. Applying the inverse power method to $\begin{bmatrix} 0 & 1 \\ -6 & 7 \end{bmatrix}$ starting at $(1,0)$, yields 0.9780 after 4 iterations. So, the root nearest 5 is $5 + (0.9780)^{-1} = 6.0225$.

39. Applying the inverse power method to $\begin{bmatrix} 0 & 1 \\ 9 & 8 \end{bmatrix}$ starting at $(1,0)$, yields -1.0006 after 4 iterations. So, the root nearest 10 is $10 + (-1.0006)^{-1} = 9.0006$.

Exercises 7.4

In Exercises 1–9 we use the vector equation

$$\begin{bmatrix} 1 \\ 4 \end{bmatrix} = \frac{3}{2}\begin{bmatrix} -1 \\ 1 \end{bmatrix} + \frac{5}{2}\begin{bmatrix} 1 \\ 1 \end{bmatrix}$$

1. (a) $\mathbf{x}_k = \frac{3}{2} \cdot 1^k \cdot \begin{bmatrix} -1 \\ 1 \end{bmatrix} + \frac{5}{2} \cdot 5^k \cdot \begin{bmatrix} 1 \\ 1 \end{bmatrix}$

(b) $A\mathbf{x}_0 = \mathbf{x}_1 = \begin{bmatrix} 11 \\ 14 \end{bmatrix}$, $A^2\mathbf{x}_0 = \mathbf{x}_2 = \begin{bmatrix} 61 \\ 64 \end{bmatrix}$

(c) Neither

3. (a) $\mathbf{x}_k = \frac{3}{2} \cdot (-7)^k \cdot \begin{bmatrix} -1 \\ 1 \end{bmatrix} + \frac{5}{2} \cdot (-1)^k \cdot \begin{bmatrix} 1 \\ 1 \end{bmatrix}$

(b) $A\mathbf{x}_0 = \mathbf{x}_1 = \begin{bmatrix} 8 \\ -13 \end{bmatrix}$, $A^2\mathbf{x}_0 = \mathbf{x}_2 = \begin{bmatrix} -71 \\ 76 \end{bmatrix}$

(c) Neither

5. (a) $\mathbf{x}_k = \frac{3}{2}\cdot\left(\frac{1}{2}\right)^k \begin{bmatrix} -1 \\ 1 \end{bmatrix} + \frac{5}{2}\cdot\left(\frac{5}{2}\right)^k \begin{bmatrix} 1 \\ 1 \end{bmatrix}$

(b) $A\mathbf{x}_0 = \mathbf{x}_1 = \begin{bmatrix} \frac{11}{2} \\ 7 \end{bmatrix}$, $A^2\mathbf{x}_0 = \mathbf{x}_2 = \begin{bmatrix} \frac{61}{4} \\ 16 \end{bmatrix}$

(c) Neither

7. (a) $\mathbf{x}_k = \frac{3}{2}\cdot(-13)^k\cdot \begin{bmatrix} -1 \\ 1 \end{bmatrix} + \frac{5}{2}\cdot(-1)^k\cdot \begin{bmatrix} 1 \\ 1 \end{bmatrix}$

(b) $A\mathbf{x}_0 = \mathbf{x}_1 = \begin{bmatrix} 17 \\ -22 \end{bmatrix}$, $A^2\mathbf{x}_0 = \mathbf{x}_2 = \begin{bmatrix} -251 \\ 256 \end{bmatrix}$

(c) Neither

9. (a) $\mathbf{x}_k = \frac{3}{2}\cdot\left(-\frac{1}{10}\right)^k \begin{bmatrix} -1 \\ 1 \end{bmatrix} + \frac{5}{2}\cdot\left(\frac{1}{2}\right)^k \begin{bmatrix} 1 \\ 1 \end{bmatrix}$

(b) $A\mathbf{x}_0 = \mathbf{x}_1 = \begin{bmatrix} \frac{7}{5} \\ \frac{11}{10} \end{bmatrix}$, $A^2\mathbf{x}_0 = \mathbf{x}_2 = \begin{bmatrix} \frac{61}{100} \\ \frac{16}{25} \end{bmatrix}$

(c) Attractor

11. (a) $\begin{bmatrix} -3 & 2 \\ 2 & -3 \end{bmatrix}^5 \begin{bmatrix} 1 \\ 1 \end{bmatrix} = \begin{bmatrix} -1 \\ -1 \end{bmatrix}$

(b) Also,

$$\mathbf{x}_5 = 0\cdot(-5)^5 \begin{bmatrix} -1 \\ 1 \end{bmatrix} + 1\cdot(-1)^5 \begin{bmatrix} 1 \\ 1 \end{bmatrix} = \begin{bmatrix} -1 \\ -1 \end{bmatrix}$$

13. (a) $\begin{bmatrix} \frac{3}{10} & \frac{2}{10} \\ \frac{2}{10} & \frac{3}{10} \end{bmatrix}^5 \begin{bmatrix} 1 \\ 1 \end{bmatrix} = \begin{bmatrix} \frac{1}{32} \\ \frac{1}{32} \end{bmatrix}$

(b) We have

$$\mathbf{x}_5 = 0\cdot\left(\frac{1}{10}\right)^5 \begin{bmatrix} -1 \\ 1 \end{bmatrix} + 1\cdot\left(\frac{1}{2}\right)^5 \begin{bmatrix} 1 \\ 1 \end{bmatrix} = \begin{bmatrix} \frac{1}{32} \\ \frac{1}{32} \end{bmatrix}$$

15. (a) $\begin{bmatrix} 5 & 4 \\ 4 & 5 \end{bmatrix}^5 \begin{bmatrix} 1 \\ 1 \end{bmatrix} = \begin{bmatrix} 59049 \\ 59049 \end{bmatrix}$

(b) We have

$$\mathbf{x}_5 = 0 \cdot 1^5 \begin{bmatrix} -1 \\ 1 \end{bmatrix} + 1 \cdot 9^5 \begin{bmatrix} 1 \\ 1 \end{bmatrix} = \begin{bmatrix} 59049 \\ 59049 \end{bmatrix}$$

17. $\mathbf{x}_1 = (0, 1)$, $\mathbf{x}_2 = \left(-\frac{1}{2}, \frac{1}{2}\right)$, $\mathbf{x}_3 = \left(-\frac{1}{2}, 0\right)$. The origin is an attractor.

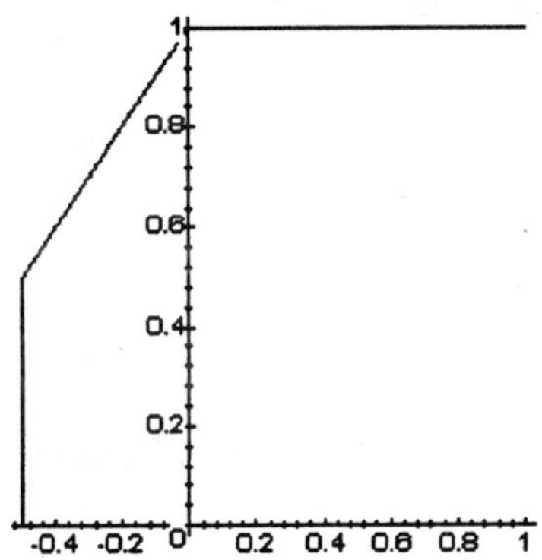

19. This is true, because both eigenvalues $\frac{1}{2} \pm \frac{\sqrt{3}}{2}i$ are 6th roots of 1. Indeed, we have $\left(\frac{1}{2} \pm \frac{\sqrt{3}}{2}i\right)^6 = 1$ and $\left(\frac{1}{2} \pm \frac{\sqrt{3}}{2}i\right)^k \neq 1$ for $k < 6$.

21. In matrix notation we have

$$\begin{bmatrix} A_{k+1} \\ B_{k+1} \end{bmatrix} = \begin{bmatrix} 0.8 & 0.7 \\ 0.2 & 0.3 \end{bmatrix}^k \begin{bmatrix} A_k \\ B_k \end{bmatrix}$$

$$\mathbf{x}_0 = \begin{bmatrix} A_0 \\ B_0 \end{bmatrix} = \begin{bmatrix} 300 \\ 100 \end{bmatrix}$$

(a) After the third generation there are about 311 females and 89 males.

(b) In the long run there are 311.11 females and 88.889 males. So, the females will eventually dominate the population.

Exercises 7.5

1. Only the second and fourth vectors are probability vectors. The first vector has entries that do not add up to 1. The third vector has a negative entry.

3. (a) The matrix is not regular because it is lower triangular hence all its powers are also lower triangular, so there are always entries that are zero.

(b) The matrix is not regular because it is upper triangular hence all its powers are also upper triangular, so there are always entries that are zero.

5. Let $A = \begin{bmatrix} \frac{1}{2} & 1 \\ \frac{1}{2} & 0 \end{bmatrix}$. Solving $[A - I : 0]$ yields $(2r, r)$. But we want $2r + r = 1$. So, $r = \frac{1}{3}$. Hence, $\mathbf{v} = \begin{bmatrix} \frac{2}{3} \\ \frac{1}{3} \end{bmatrix}$ is the steady state vector of A.

Let $B = \begin{bmatrix} 0 & \frac{1}{2} \\ 1 & \frac{1}{2} \end{bmatrix}$. Solving $[B - I : 0]$ yields $\left(\frac{r}{2}, r\right)$. But we want $\frac{r}{2} + r = 1$. So, $r = \frac{2}{3}$. Therefore, $\mathbf{v} = \begin{bmatrix} \frac{1}{3} \\ \frac{2}{3} \end{bmatrix}$ is the steady state vector of B.

7. (a) The solution of $[A - I : 0]$ is obtained from

$$\begin{bmatrix} \frac{1}{5} - 1 & \frac{1}{2} & \frac{1}{2} & 0 \\ \frac{1}{5} & \frac{1}{2} - 1 & 0 & 0 \\ \frac{3}{5} & 0 & \frac{1}{2} - 1 & 0 \end{bmatrix} \sim \begin{bmatrix} 1 & 0 & -\frac{5}{6} & 0 \\ 0 & 1 & -\frac{1}{3} & 0 \\ 0 & 0 & 0 & 0 \end{bmatrix}$$

and is $\left(\frac{5}{6}r, \frac{1}{3}r, r\right)$. But $\frac{5}{6}r + \frac{1}{3}r + r = 1$, so $r = \frac{6}{13}$. Therefore, the steady-state vector is $\left(\frac{5}{13}, \frac{2}{13}, \frac{6}{13}\right)$.

(b) The solution of $[A - I : 0]$ is obtained from

$$\begin{bmatrix} \frac{2}{5} - 1 & \frac{1}{5} & \frac{1}{5} & 0 \\ \frac{2}{5} & \frac{2}{5} - 1 & \frac{3}{10} & 0 \\ \frac{1}{5} & \frac{2}{5} & \frac{1}{2} - 1 & 0 \end{bmatrix} \sim \begin{bmatrix} 1 & 0 & -\frac{9}{14} & 0 \\ 0 & 1 & -\frac{13}{14} & 0 \\ 0 & 0 & 0 & 0 \end{bmatrix}$$

and is $\left(\frac{9}{14}r, \frac{13}{14}r, r\right)$. But $\frac{9}{14}r + \frac{13}{14}r + r = 1$,so $r = \frac{7}{18}$. Therefore, the steady-state vector is $\left(\frac{1}{4}, \frac{13}{36}, \frac{7}{18}\right)$.

9. To show that the steady state vector $\mathbf{v}$ is unique, let us assume that $\mathbf{u}$ is another probability vector such that $A\mathbf{u} = \mathbf{u}$. Then from the first part of the theorem: $\lim_{k\to\infty} A^k\mathbf{u} = \mathbf{v}$. Because $A\mathbf{u} = \mathbf{u}$, we have $A^k\mathbf{u} = \mathbf{u}$ for all k. Hence, $\mathbf{u} = \mathbf{v}$.

11. M is regular because M^2 has only positive entries. The equilibrium of M (calculated in Exercise 7) is $(\frac{5}{13}, \frac{2}{13}, \frac{6}{13})$. So, in the long run plan C is the most popular and plan B is the least popular.

Exercises 8.1

1. The set is orthogonal because

$$(1,-2,1)\cdot(4,2,0)=0 \quad (1,-2,1)\cdot(-1,2,5)=0 \quad (4,2,0)\cdot(-1,2,5)=0$$

It also forms an orthogonal basis of $\mathbf{R}^3$.

3. The set is orthogonal because

$$(1,1,-1,1)\cdot(1,1,1,-1)=0,\ (1,1,-1,1)\cdot(0,0,1,1)=0,\ (1,1,1,-1)\cdot(0,0,1,1)=0$$

This set does not form an orthogonal basis of $\mathbf{R}^4$. Three vectors do not span $\mathbf{R}^4$.

5. $\mathbf{v}_1=(1,1,2)$, $\mathbf{v}_2=(-1,1,0)$, $\mathbf{v}_3=(0,0,1)$

7. Every pair of vectors has dot product zero hence, the vectors are linearly independent. Thus, they form an orthogonal basis of $\mathbf{R}^3$. We have

$$(1,1,1)=\frac{3}{19}\mathbf{v}_1+\frac{1}{19}\mathbf{v}_2+1\mathbf{v}_3$$

9. Every pair of vectors has dot product zero hence, the vectors are linearly independent. Thus, they form an orthogonal basis of $\mathbf{R}^3$. We have

$$(1,1,1)=0\mathbf{v}_1+\frac{1}{7}\mathbf{v}_2+\frac{1}{7}\mathbf{v}_3$$

11. The vectors are orthogonal but not orthonormal. Diving each vector by its length yields the orthonormal vectors $\begin{bmatrix}\frac{1}{\sqrt{5}}\\ \frac{2}{\sqrt{5}}\end{bmatrix}, \begin{bmatrix}-\frac{2}{\sqrt{5}}\\ \frac{1}{\sqrt{5}}\end{bmatrix}$.

13. The vectors are orthonormal. This can be seen from the calculation

$$\begin{bmatrix}\frac{1}{3} & \frac{2}{3} & \frac{2}{3}\\ \frac{2}{3} & -\frac{2}{3} & \frac{1}{3}\\ \frac{2}{3} & \frac{1}{3} & -\frac{2}{3}\end{bmatrix}^T \begin{bmatrix}\frac{1}{3} & \frac{2}{3} & \frac{2}{3}\\ \frac{2}{3} & -\frac{2}{3} & \frac{1}{3}\\ \frac{2}{3} & \frac{1}{3} & -\frac{2}{3}\end{bmatrix} = \begin{bmatrix}1 & 0 & 0\\ 0 & 1 & 0\\ 0 & 0 & 1\end{bmatrix}$$

15. $\mathcal{B}$ is an orthonormal basis for $\mathbf{R}^2$ because both vectors are unit and their dot product is zero. We have

$$\mathbf{e}_1 = \frac{2}{\sqrt{5}}\mathbf{v}_1 - \frac{1}{\sqrt{5}}\mathbf{v}_2$$

17. The matrix is orthogonal. Its inverse is $\begin{bmatrix} 0 & -1 \\ 1 & 0 \end{bmatrix}$.

19. The matrix is orthogonal (it is easy to check that $A^T A = I$). Its inverse is

$$\begin{bmatrix} \frac{3}{\sqrt{14}} & -\frac{2}{\sqrt{14}} & \frac{1}{\sqrt{14}} \\ \frac{1}{\sqrt{6}} & \frac{2}{\sqrt{6}} & \frac{1}{\sqrt{6}} \\ -\frac{2}{\sqrt{21}} & -\frac{1}{\sqrt{21}} & \frac{4}{\sqrt{21}} \end{bmatrix}$$

21. If the columns of A form an orthonormal set, then they are n linearly independent m-vectors. Therefore, $m \geq n$. If A only had orthogonal columns, then some columns could be zero so the assumption $m \geq n$ is not necessarily true.

23. If $\mathcal{B} = \{\mathbf{v}_1, \ldots, \mathbf{v}_n\}$, then

$$\mathbf{u} = \frac{\mathbf{u}\cdot\mathbf{v}_1}{\mathbf{v}_1\cdot\mathbf{v}_1}\mathbf{v}_1 + \cdots + \frac{\mathbf{u}\cdot\mathbf{v}_n}{\mathbf{v}_n\cdot\mathbf{v}_n}\mathbf{v}_n = \frac{0}{\mathbf{v}_1\cdot\mathbf{v}_1}\mathbf{v}_1 + \cdots + \frac{0}{\mathbf{v}_n\cdot\mathbf{v}_n}\mathbf{v}_n = \mathbf{0}$$

25. $2 \Rightarrow 3$. Let us assume that $A\mathbf{u}\cdot A\mathbf{v} = \mathbf{u}\cdot\mathbf{v}$ for any n-vectors $\mathbf{u}$ and $\mathbf{v}$. In particular, $A\mathbf{v}\cdot A\mathbf{v} = \mathbf{v}\cdot\mathbf{v}$, or $\|A\mathbf{v}\|^2 = \|\mathbf{v}\|^2$. Hence, $\|A\mathbf{v}\| = \|\mathbf{v}\|$.

$3 \Rightarrow 2$. Let us now assume that $\|A\mathbf{v}\| = \|\mathbf{v}\|$ for any vector $\mathbf{v}$. Therefore, for any two vectors $\mathbf{u}$ and $\mathbf{v}$ we have $\|A(\mathbf{u}-\mathbf{v})\| = \|\mathbf{u}-\mathbf{v}\|$. So, $\|A\mathbf{u} - A\mathbf{v}\|^2 = \|\mathbf{u}-\mathbf{v}\|^2$. Hence,

$$(A\mathbf{u} - A\mathbf{v})\cdot(A\mathbf{u} - A\mathbf{v}) = (\mathbf{u}-\mathbf{v})\cdot(\mathbf{u}-\mathbf{v})$$

We expand to get

$$A\mathbf{u}\cdot A\mathbf{u} - 2A\mathbf{u}\cdot A\mathbf{v} + A\mathbf{v}\cdot A\mathbf{v} = \mathbf{u}\cdot\mathbf{u} - 2\mathbf{u}\cdot\mathbf{v} + \mathbf{v}\cdot\mathbf{v}$$

But $A\mathbf{u}\cdot A\mathbf{u} = \mathbf{u}\cdot\mathbf{u}$ and $A\mathbf{v}\cdot A\mathbf{v} = \mathbf{v}\cdot\mathbf{v}$, because $\|A\mathbf{u}\| = \|\mathbf{u}\|$, $\|A\mathbf{v}\| = \|\mathbf{v}\|$. Therefore,

$$A\mathbf{u}\cdot A\mathbf{v} = \mathbf{u}\cdot\mathbf{v}$$

27. Let $\mathbf{v} = (\mathbf{u}\cdot\mathbf{v}_1)\mathbf{v}_1 + \cdots + (\mathbf{u}\cdot\mathbf{v}_k)\mathbf{v}_k$. We have

$$\begin{aligned}\mathbf{u}\cdot\mathbf{v} &= ((\mathbf{u}\cdot\mathbf{v}_1)\mathbf{v}_1 + \cdots + (\mathbf{u}\cdot\mathbf{v}_k)\mathbf{v}_k)\cdot\mathbf{u} \\ &= (\mathbf{u}\cdot\mathbf{v}_1)(\mathbf{v}_1\cdot\mathbf{u}) + \cdots + (\mathbf{u}\cdot\mathbf{v}_k)(\mathbf{v}_k\cdot\mathbf{u}) \\ &= (\mathbf{u}\cdot\mathbf{v}_1)^2 + \cdots + (\mathbf{u}\cdot\mathbf{v}_k)^2\end{aligned}$$

Also,

$$\begin{aligned}\mathbf{v}\cdot\mathbf{v} &= (\textstyle\sum_i(\mathbf{u}\cdot\mathbf{v}_i)\mathbf{v}_i)\cdot\left(\textstyle\sum_j(\mathbf{u}\cdot\mathbf{v}_j)\mathbf{v}_j\right) \\ &= \textstyle\sum_{i,j}(\mathbf{u}\cdot\mathbf{v}_i)(\mathbf{u}\cdot\mathbf{v}_j)(\mathbf{v}_i\cdot\mathbf{v}_j) \\ &= \textstyle\sum_i(\mathbf{u}\cdot\mathbf{v}_i)(\mathbf{u}\cdot\mathbf{v}_i) \\ &= (\mathbf{u}\cdot\mathbf{v}_1)^2 + \cdots + (\mathbf{u}\cdot\mathbf{v}_k)^2\end{aligned}$$

because $\mathbf{v}_i\cdot\mathbf{v}_j$ is 0, for $i \neq j$ and 1, for $i = j$. We conclude that

$$\mathbf{u}\cdot\mathbf{v} = \mathbf{v}\cdot\mathbf{v}$$

Now let $\mathbf{r} = \mathbf{u} - \mathbf{v}$. We show that $\mathbf{v}$ and $\mathbf{r}$ are orthogonal.

$$\mathbf{r}\cdot\mathbf{v} = (\mathbf{u}-\mathbf{v})\cdot\mathbf{v} = \mathbf{u}\cdot\mathbf{v} - \mathbf{v}\cdot\mathbf{v} = 0$$

Hence, by the Pythagorean Theorem for n-vectors we have

$$\|\mathbf{u}\|^2 = \|\mathbf{v}\|^2 + \|\mathbf{r}\|^2$$

and because $\|\mathbf{r}\|^2 \geq 0$ we conclude that $\|\mathbf{u}\|^2 \geq \|\mathbf{v}\|^2 = \mathbf{v}\cdot\mathbf{v}$. Therefore,

$$\|\mathbf{u}\|^2 \geq (\mathbf{u}\cdot\mathbf{v}_1)^2 + \cdots + (\mathbf{u}\cdot\mathbf{v}_k)^2$$

This proves Bessel's inequality.

Exercises 8.2

1. The projection is zero, so $\mathbf{u}$ is on the line.

$$\mathbf{u}_{\text{pr}} = \frac{(-2,1)\cdot(1,2)}{(1,2)\cdot(1,2)}(1,2) = (0,0)$$

3. We have

$$\mathbf{u}_{\text{pr}} = \frac{(3,-1,2)\cdot(1,1,-3)}{(1,1,-3)\cdot(1,1,-3)}(1,1,-3) = \left(-\frac{4}{11}, -\frac{4}{11}, \frac{12}{11}\right)$$

5. The orthogonal projection is $\mathbf{u}_{\text{pr}} = \frac{(-2,1)\cdot(2,3)}{(2,3)\cdot(2,3)}(2,3) = \left(-\frac{2}{13}, -\frac{3}{13}\right)$. Therefore, $\mathbf{u} = \left(-\frac{2}{13}, -\frac{3}{13}\right) + \left(-\frac{24}{13}, \frac{16}{13}\right)$.

7. Because $\mathbf{u}_{\text{pr}} = \frac{(1,3)\cdot(-2,2)}{(-2,2)\cdot(-2,2)}(-2,2) = (-1,1)$, we get $\mathbf{u} = (-1,1) + (2,2)$.

9. We have

$$\begin{aligned}\mathbf{u}_{\text{pr}} &= \frac{(1,1,1)\cdot(1,-1,2)}{(1,-1,2)\cdot(1,-1,2)}(1,-1,2) + \frac{(1,1,1)\cdot(-2,2,2)}{(-2,2,2)\cdot(-2,2,2)}(-2,2,2) \\ &= (0,0,1)\end{aligned}$$

11. $\left(V^{\perp}\right)^{\perp}$ consists of all n-vectors orthogonal to $V^{\perp}$ which is V. Thus, $\left(V^{\perp}\right)^{\perp} = V$.

13. We have

$$\begin{aligned}\mathbf{u}_1 &= (2,-1,1) \\ \mathbf{u}_2 &= (0,3,-1) - \frac{(0,3,-1)\cdot(2,-1,1)}{(2,-1,1)\cdot(2,-1,1)}(2,-1,1) = \left(\frac{4}{3}, \frac{7}{3}, -\frac{1}{3}\right) \\ \mathbf{u}_3 &= (1,2,0) - \frac{(1,2,0)\cdot(2,-1,1)}{(2,-1,1)\cdot(2,-1,1)}(2,-1,1) - \frac{(1,2,0)\cdot\left(\frac{4}{3},\frac{7}{3},-\frac{1}{3}\right)}{\left(\frac{4}{3},\frac{7}{3},-\frac{1}{3}\right)\cdot\left(\frac{4}{3},\frac{7}{3},-\frac{1}{3}\right)}\left(\frac{4}{3},\frac{7}{3},-\frac{1}{3}\right) \\ &= \left(-\frac{1}{11}, \frac{1}{11}, \frac{3}{11}\right)\end{aligned}$$

Thus, we get the following orthogonal basis of $\mathbf{R}^3$.

$$\left\{ \begin{bmatrix} 2 \\ -1 \\ 1 \end{bmatrix}, \begin{bmatrix} \frac{4}{3} \\ \frac{7}{3} \\ -\frac{1}{3} \end{bmatrix}, \begin{bmatrix} -\frac{1}{11} \\ \frac{1}{11} \\ \frac{3}{11} \end{bmatrix} \right\}$$

Normalization yields the orthonormal basis:

$$\left\{ \begin{bmatrix} \frac{2}{\sqrt{6}} \\ \frac{-1}{\sqrt{6}} \\ \frac{1}{\sqrt{6}} \end{bmatrix}, \begin{bmatrix} \frac{4}{\sqrt{66}} \\ \frac{7}{\sqrt{66}} \\ -\frac{1}{\sqrt{66}} \end{bmatrix}, \begin{bmatrix} -\frac{1}{\sqrt{11}} \\ \frac{1}{\sqrt{11}} \\ \frac{3}{\sqrt{11}} \end{bmatrix} \right\}$$

15. We let $\mathbf{u}_1 = (4, 2, -1)$ and let

$$\mathbf{u}_2 = (1, 2, 3) - \frac{(1, 2, 3) \cdot (4, 2, -1)}{(4, 2, -1) \cdot (4, 2, -1)}(4, 2, -1) = \left(\frac{1}{21}, \frac{32}{21}, \frac{68}{21}\right)$$

So, we get $\left\{(4, 2, -1), \left(\frac{1}{21}, \frac{32}{21}, \frac{68}{21}\right)\right\}$.

17. First we apply Gram–Schmidt

$$\begin{aligned} \mathbf{u}_1 &= (5, 1, -4) \\ \mathbf{u}_2 &= (0, 2, 1) - \frac{(0, 2, 1) \cdot (5, 1, -4)}{(5, 1, -4) \cdot (5, 1, -4)}(5, 1, -4) = \left(\frac{5}{21}, \frac{43}{21}, \frac{17}{21}\right) \end{aligned}$$

to get an orthogonal basis for the span $\left\{(5, 1, -4), \left(\frac{5}{21}, \frac{43}{21}, \frac{17}{21}\right)\right\}$. Then we compute the projection $\mathbf{u}_{\mathbf{pr}}$ onto V

$$\begin{aligned} \mathbf{u}_{\mathbf{pr}} &= \frac{(2, 0, 1) \cdot (5, 1, -4)}{(5, 1, -4) \cdot (5, 1, -4)}(5, 1, -4) + \frac{(2, 0, 1) \cdot (\frac{5}{21}, \frac{43}{21}, \frac{17}{21})}{(\frac{5}{21}, \frac{43}{21}, \frac{17}{21}) \cdot (\frac{5}{21}, \frac{43}{21}, \frac{17}{21})}\left(\frac{5}{21}, \frac{43}{21}, \frac{17}{21}\right) \\ &= \left(\frac{80}{103}, \frac{70}{103}, -\frac{37}{103}\right) \end{aligned}$$

Thus, $\mathbf{u} = \left(\frac{80}{103}, \frac{70}{103}, -\frac{37}{103}\right) + \left(\frac{126}{103}, -\frac{70}{103}, \frac{140}{103}\right)$.

19. The projection of $\mathbf{u}$ onto V is

$$\begin{aligned}\mathbf{u}_{\mathrm{pr}} &= \frac{(4,-4,4)\cdot(1,1,-2)}{(1,1,-2)\cdot(1,1,-2)}(1,1,-2)+\frac{(4,-4,4)\cdot(2,0,1)}{(2,0,1)\cdot(2,0,1)}(2,0,1)\\ &= \left(\frac{52}{15},-\frac{4}{3},\frac{76}{15}\right)\end{aligned}$$

Hence, $\mathbf{u}_{\mathrm{c}} = (4,-4,4) - \left(\frac{52}{15},-\frac{4}{3},\frac{76}{15}\right) = \left(\frac{8}{15},-\frac{8}{3},-\frac{16}{15}\right)$. So the distance from $\mathbf{u}$ to V is

$$\|\mathbf{u}_{\mathrm{c}}\| = \left\|\left(\frac{8}{15},-\frac{8}{3},-\frac{16}{15}\right)\right\| = \frac{8}{15}\sqrt{30}$$

21. The projection of $\mathbf{u}$ onto V is

$$\begin{aligned}\mathbf{u}_{\mathrm{pr}} &= \frac{(1,2,1,2)\cdot(1,1,-2,2)}{(1,1,-2,2)\cdot(1,1,-2,2)}(1,1,-2,2)+\frac{(1,2,1,2)\cdot(2,0,1,0)}{(2,0,1,0)\cdot(2,0,1,0)}(2,0,1,0)\\ &= \left(\frac{17}{10},\frac{1}{2},-\frac{2}{5},1\right)\end{aligned}$$

Hence, $\mathbf{u}_{\mathrm{c}} = (1,2,1,2) - \left(\frac{17}{10},\frac{1}{2},-\frac{2}{5},1\right) = \left(-\frac{7}{10},\frac{3}{2},\frac{7}{5},1\right)$. So the distance from $\mathbf{u}$ to V is

$$\|\mathbf{u}_{\mathrm{c}}\| = \left\|\left(-\frac{7}{10},\frac{3}{2},\frac{7}{5},1\right)\right\| = \frac{1}{10}\sqrt{570}$$

Exercises 8.3

---

1. $\begin{bmatrix} 0 & 1 \\ -1 & 0 \end{bmatrix}\begin{bmatrix} -1 & -3 \\ 0 & -2 \end{bmatrix}$

3. $\begin{bmatrix} 0 & 0 & -1 \\ 0 & -1 & 0 \\ 1 & 0 & 0 \end{bmatrix}\begin{bmatrix} -1 & 0 & 2 \\ 0 & -1 & -2 \\ 0 & 0 & -4 \end{bmatrix}$

5. $\begin{bmatrix} \frac{\sqrt{2}}{2} & -\frac{\sqrt{3}}{3} & \frac{\sqrt{6}}{6} \\ 0 & \frac{\sqrt{3}}{3} & \frac{\sqrt{6}}{3} \\ -\frac{\sqrt{2}}{2} & -\frac{\sqrt{3}}{3} & \frac{\sqrt{6}}{6} \end{bmatrix}\begin{bmatrix} \sqrt{2} & -\sqrt{2} & 0 \\ 0 & \sqrt{3} & 0 \\ 0 & 0 & \sqrt{6} \end{bmatrix}$

7. $\begin{bmatrix} \frac{1}{2} & -\frac{\sqrt{5}}{10} \\ \frac{1}{2} & \frac{\sqrt{5}}{10} \\ \frac{1}{2} & \frac{3\sqrt{5}}{10} \\ -\frac{1}{2} & \frac{3\sqrt{5}}{10} \end{bmatrix} \begin{bmatrix} 2 & -2 \\ 0 & 2\sqrt{5} \end{bmatrix}$

9. $\begin{bmatrix} 2 & 0 \\ 0 & \sqrt{6} \end{bmatrix}$

11. Let $Q = A$. Such Q is acceptable because A has orthogonal columns. Then $R = Q^T A = A^T A = I$, because A is orthonormal. So, $A = AI$ is a QR factorization of A.

13. Let $A\mathbf{x} = \mathbf{b}$ be consistent and let $\mathbf{v}$ be a solution. So, $A\mathbf{v} = \mathbf{b}$. Hence, $Q(R\mathbf{v}) = \mathbf{b}$. Therefore, the system $Q\mathbf{y} = \mathbf{b}$ is consistent because $R\mathbf{v}$ is a solution of it.

Conversely, let $Q\mathbf{y} = \mathbf{b}$ be consistent. Let $\mathbf{u}$ be a solution. Then $Q\mathbf{u} = \mathbf{b}$. Because, R is invertible we may define $\mathbf{v} = R^{-1}\mathbf{u}$. Hence, $\mathbf{u} = R\mathbf{v}$ and $Q\mathbf{u} = QR\mathbf{v} = \mathbf{b}$. So, $A\mathbf{v} = \mathbf{b}$. Therefore, $\mathbf{v}$ is a solution of $A\mathbf{x} = \mathbf{b}$, so this system is consistent.

Next we show that A and Q have the same column space. Let $\mathbf{b}$ be in the column space of A. Hence, the system $A\mathbf{x} = \mathbf{b}$ is consistent. Therefore $Q\mathbf{y} = \mathbf{b}$ is consistent, so $\mathbf{b}$ is in the column space of Q. If on the other hand, $\mathbf{b}$ is in the column space of Q, then $Q\mathbf{y} = \mathbf{b}$ is consistent, so $A\mathbf{x} = \mathbf{b}$ is consistent. Hence, $\mathbf{b}$ is in the column space of A. We conclude that A and Q have the same column space.

15. Because

$$A_1 = \begin{bmatrix} 8.8916 & 6.1382 \\ 0.13817 & 1.1075 \end{bmatrix}, \quad A_2 = \begin{bmatrix} 8.9869 & 6.0159 \\ 0.015947 & 1.0121 \end{bmatrix},$$

$$A_3 = \begin{bmatrix} 8.9970 & 6.0013 \\ 0.0015697 & 1.0013 \end{bmatrix}$$

the estimated eigenvalues are 8.9970, 1.0013. The true eigenvalues are 9, 1. The errors are 0.003, −0.0013.

17. Because

$$A_1 = \begin{bmatrix} 13.864 & 9.1909 \\ 0.19135 & 1.1367 \end{bmatrix}, \quad A_2 = \begin{bmatrix} 13.992 & 9.0157 \\ 0.015813 & 1.0109 \end{bmatrix},$$

$$A_3 = \begin{bmatrix} 14.002 & 8.9943 \\ -0.0054328 & 0.99650 \end{bmatrix}$$

the estimated eigenvalues are 14.002, 0.99650. The true eigenvalues are $14, 1$. The errors are $-0.002, 0.0035$.

Exercises 8.4

---

1. The normal equations $\begin{bmatrix} 3 & -1 \\ -1 & 9 \end{bmatrix} \tilde{\mathbf{x}} = \begin{bmatrix} -2 \\ 6 \end{bmatrix}$ yield $\tilde{\mathbf{x}} = \left(-\frac{6}{13}, \frac{8}{13}\right)$.

3. The normal equations $\begin{bmatrix} 10 & 5 \\ 5 & 6 \end{bmatrix} \tilde{\mathbf{x}} = \begin{bmatrix} -1 \\ 3 \end{bmatrix}$ imply $\tilde{\mathbf{x}} = \left(-\frac{3}{5}, 1\right)$.

5. All we need is a system with invertible coefficient matrix. For example:

$$\begin{bmatrix} 2 & 1 & 0 \\ 0 & 1 & 1 \\ 0 & 0 & -3 \end{bmatrix} \mathbf{x} = \begin{bmatrix} 9 \\ 6 \\ 3 \end{bmatrix}$$

with solution $(1, 7, -1)$, by direct elimination, or by the normal equations.

7. The normal equations

$$\begin{bmatrix} 1 & -1 \\ 1 & 0 \\ 1 & 1 \end{bmatrix}^T \begin{bmatrix} 1 & -1 \\ 1 & 0 \\ 1 & 1 \end{bmatrix} = \begin{bmatrix} 1 & -1 \\ 1 & 0 \\ 1 & 1 \end{bmatrix}^T \begin{bmatrix} -2 \\ 1 \\ 5 \end{bmatrix}$$

simplify to the system

$$\begin{bmatrix} 3 & 0 \\ 0 & 2 \end{bmatrix} \tilde{\mathbf{x}} = \begin{bmatrix} 4 \\ 7 \end{bmatrix}$$

with solution $\tilde{\mathbf{x}} = \left(\frac{4}{3}, \frac{7}{2}\right)$. Therefore, the equation of the line is $y = \frac{7}{2}x + \frac{4}{3}$.

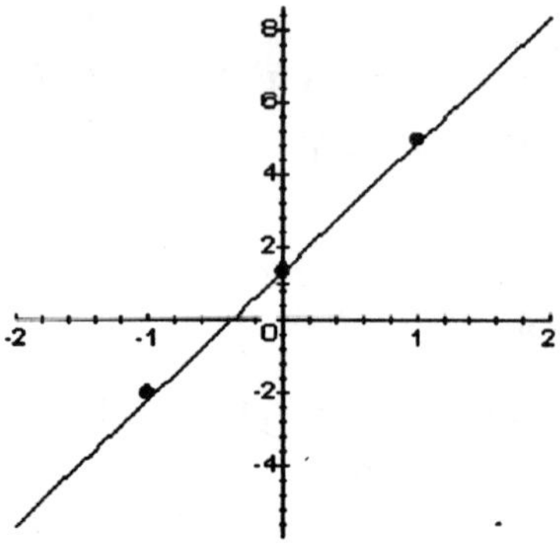

9. The normal equations

$$\begin{bmatrix} 1 & -1 \\ 1 & 1 \\ 1 & 3 \\ 1 & 4 \end{bmatrix}^T \begin{bmatrix} 1 & -1 \\ 1 & 1 \\ 1 & 3 \\ 1 & 4 \end{bmatrix} = \begin{bmatrix} 1 & -1 \\ 1 & 1 \\ 1 & 3 \\ 1 & 4 \end{bmatrix}^T \begin{bmatrix} -1 \\ 2 \\ 2 \\ 4 \end{bmatrix}$$

simplify to the system

$$\begin{bmatrix} 4 & 7 \\ 7 & 27 \end{bmatrix} \widetilde{\mathbf{x}} = \begin{bmatrix} 7 \\ 25 \end{bmatrix}$$

with solution $\widetilde{\mathbf{x}} = \left(\frac{14}{59}, \frac{51}{59}\right)$. Therefore, the equation of the line is $y = \frac{51}{59}x + \frac{14}{59}$.

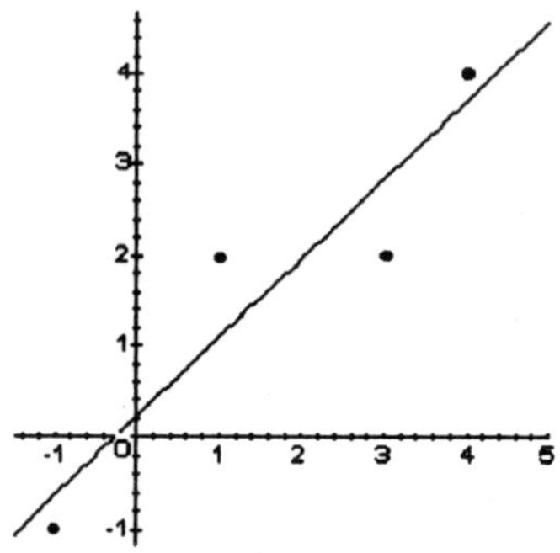

11. We solve the normal equations

$$\begin{bmatrix} 5 & 1 & -3 \\ 1 & 1 & 1 \\ -3 & 1 & 5 \end{bmatrix} \widetilde{\mathbf{x}} = \begin{bmatrix} -2 \\ 2 \\ 6 \end{bmatrix}$$

by reducing

$$\begin{bmatrix} 5 & 1 & -3 & -2 \\ 1 & 1 & 1 & 2 \\ -3 & 1 & 5 & 6 \end{bmatrix} \sim \begin{bmatrix} 1 & 0 & -1 & -1 \\ 0 & 1 & 2 & 3 \\ 0 & 0 & 0 & 0 \end{bmatrix}$$

We get $\widetilde{\mathbf{x}} = (r-1, -2r+3, r)$, $r \in \mathbf{R}$.

13. From the system $R\widetilde{\mathbf{x}} = Q^T\mathbf{b}$ we get

$$\begin{bmatrix} 5 & 10 \\ 0 & 5 \end{bmatrix} \widetilde{\mathbf{x}} = \begin{bmatrix} -4 \\ \frac{9}{5} \end{bmatrix} \Rightarrow \widetilde{\mathbf{x}} = \left(-\frac{38}{25}, \frac{9}{25}\right)$$

15. Because A has orthogonal columns, A^TA is diagonal with diagonal entries $\|\mathbf{a}_1\|^2$, $\ldots$, $\|\mathbf{a}_n\|^2$, by Theorem 4, Section 8.1. Also, $A^T\mathbf{b}$ has components $\mathbf{b}\cdot\mathbf{a}_1, \ldots, \mathbf{b}\cdot\mathbf{a}_n$. Hence, the normal equations

$$A^TA\widetilde{\mathbf{x}} = A^T\mathbf{b}$$

take the form

$$\begin{bmatrix} \mathbf{a}_1\cdot\mathbf{a}_1 & \cdots & 0 \\ \vdots & \ddots & \vdots \\ 0 & \cdots & \mathbf{a}_n\cdot\mathbf{a}_n \end{bmatrix} \widetilde{\mathbf{x}} = \begin{bmatrix} \mathbf{b}\cdot\mathbf{a}_1 \\ \vdots \\ \mathbf{b}\cdot\mathbf{a}_n \end{bmatrix}$$

Therefore,

$$\widetilde{\mathbf{x}} = \begin{bmatrix} \dfrac{\mathbf{b}\cdot\mathbf{a}_1}{\mathbf{a}_1\cdot\mathbf{a}_1} \\ \vdots \\ \dfrac{\mathbf{b}\cdot\mathbf{a}_n}{\mathbf{a}_n\cdot\mathbf{a}_n} \end{bmatrix}$$

17. By Exercise 15, we have $\widetilde{\mathbf{x}} = \left(\frac{5}{9}, -\frac{1}{9}\right)$.

19. By Exercise 15, we have $\widetilde{\mathbf{x}} = \left(\frac{5}{3}, -\frac{1}{3}\right)$.

Exercises 8.5

1. The matrix is not symmetric.

3. The matrix is not symmetric.

5. For example, we may take $A = \begin{bmatrix} \frac{1}{\sqrt{2}} & \frac{1}{\sqrt{2}} \\ \frac{1}{\sqrt{2}} & -\frac{1}{\sqrt{2}} \end{bmatrix}$. We have $A^2 = \begin{bmatrix} 1 & 0 \\ 0 & 1 \end{bmatrix}$.

7. $A^{-1} = A = \begin{bmatrix} \frac{3}{5} & \frac{4}{5} \\ \frac{4}{5} & -\frac{3}{5} \end{bmatrix}$, by Exercise 6.

9. The eigenvalues are $1, -3$ with correspoding basic eigenvectors $(1,1)$ and $(-1,1)$, which we normalize to get

$$Q = \begin{bmatrix} \frac{1}{\sqrt{2}} & -\frac{1}{\sqrt{2}} \\ \frac{1}{\sqrt{2}} & \frac{1}{\sqrt{2}} \end{bmatrix}, \qquad D = \begin{bmatrix} 1 & 0 \\ 0 & -3 \end{bmatrix}$$

11. The eigenvalues are $2, 10$ with correspoding basic eigenvectors $(1,1)$ and $(-1,1)$, which we normalize to get

$$Q = \begin{bmatrix} \frac{1}{\sqrt{2}} & -\frac{1}{\sqrt{2}} \\ \frac{1}{\sqrt{2}} & \frac{1}{\sqrt{2}} \end{bmatrix}, \qquad D = \begin{bmatrix} 2 & 0 \\ 0 & 10 \end{bmatrix}$$

13. The eigenvalues are $0, 1, 4$ with correspoding basic eigenvectors $(0,1,1)$, $(1,0,0)$, and $(0,-1,1)$. We normalize to get

$$Q = \begin{bmatrix} 0 & 1 & 0 \\ \frac{1}{\sqrt{2}} & 0 & -\frac{1}{\sqrt{2}} \\ \frac{1}{\sqrt{2}} & 0 & \frac{1}{\sqrt{2}} \end{bmatrix}, \qquad D = \begin{bmatrix} 0 & 0 & 0 \\ 0 & 1 & 0 \\ 0 & 0 & 4 \end{bmatrix}$$

15. The eigenvalues are $2, 4, 0, 0$ with correspoding basic eigenvectors $(-1,1,0,0)$, $(0,0,-1,1)$, $(0,0,1,1)$, and $(1,1,0,0)$. We normalize to get

$$Q = \begin{bmatrix} -\frac{1}{\sqrt{2}} & 0 & 0 & \frac{1}{\sqrt{2}} \\ \frac{1}{\sqrt{2}} & 0 & 0 & \frac{1}{\sqrt{2}} \\ 0 & -\frac{1}{\sqrt{2}} & \frac{1}{\sqrt{2}} & 0 \\ 0 & \frac{1}{\sqrt{2}} & \frac{1}{\sqrt{2}} & 0 \end{bmatrix}, \qquad D = \begin{bmatrix} 2 & 0 & 0 & 0 \\ 0 & 4 & 0 & 0 \\ 0 & 0 & 0 & 0 \\ 0 & 0 & 0 & 0 \end{bmatrix}$$

17. A is symmetric, hence, it is orthogonally diagonalizable, by the spectral theorem. So, there are Q orthogonal and D diagonal such that

$$Q^T AQ = D$$

Therefore,

$$Q^T A^2 Q = (Q^T AQ)(Q^T AQ) = DD = D^2$$

Thus, Q and D^2 orthogonally diagonalize A^2 and because the eigenvalues of A^2 are the diagonal entries of D^2 they must be nonnegative.

19. $Q = \begin{bmatrix} -\frac{1}{\sqrt{2}} & \frac{1}{\sqrt{2}} \\ \frac{1}{\sqrt{2}} & \frac{1}{\sqrt{2}} \end{bmatrix}, \quad T = \begin{bmatrix} 1 & -6 \\ 0 & 11 \end{bmatrix}.$

21. $Q = \begin{bmatrix} -\frac{1}{\sqrt{2}} & \frac{1}{\sqrt{2}} \\ \frac{1}{\sqrt{2}} & \frac{1}{\sqrt{2}} \end{bmatrix}, \quad T = \begin{bmatrix} 1 & -7 \\ 0 & 12 \end{bmatrix}.$

23. We have

$$\begin{aligned} A &= QDQ^T \\ &= [\mathbf{v}_1 \cdots \mathbf{v}_n] \begin{bmatrix} \lambda_1 & \cdots & 0 \\ \vdots & \ddots & \vdots \\ 0 & \cdots & \lambda_n \end{bmatrix} \begin{bmatrix} \mathbf{v}_1 \\ \vdots \\ \mathbf{v}_n \end{bmatrix}^T \\ &= [\lambda_1 \mathbf{v}_1 \cdots \lambda_n \mathbf{v}_n] \begin{bmatrix} \mathbf{v}_1 \\ \vdots \\ \mathbf{v}_n \end{bmatrix}^T . \\ &= \lambda_1 \mathbf{v}_1 \mathbf{v}_1^T + \cdots + \lambda_n \mathbf{v}_n \mathbf{v}_n^T \end{aligned}$$

25. The spectral decomposition of A is

$$5 \begin{bmatrix} \frac{1}{\sqrt{5}} \\ -\frac{2}{\sqrt{5}} \end{bmatrix} \begin{bmatrix} \frac{1}{\sqrt{5}} \\ -\frac{2}{\sqrt{5}} \end{bmatrix}^T + 15 \begin{bmatrix} \frac{2}{\sqrt{5}} \\ \frac{1}{\sqrt{5}} \end{bmatrix} \begin{bmatrix} \frac{2}{\sqrt{5}} \\ \frac{1}{\sqrt{5}} \end{bmatrix}^T$$

Exercises 8.6

---

1. $q(\mathbf{x}) = -2x^2 + 3y^2 + 4xy$

3. $q(\mathbf{x}) = x^2 + 5z^2 - 6xy + 4xz + 2yz$

5. $\begin{bmatrix} 3 & -3 \\ -3 & 3 \end{bmatrix}$

7. $\begin{bmatrix} -4 & 1 \\ 1 & -4 \end{bmatrix}$

9. $\begin{bmatrix} 2 & 0 & 1 \\ 0 & 0 & 0 \\ 1 & 0 & 2 \end{bmatrix}$

11. $\begin{bmatrix} 5 & -4 & 0 \\ -4 & 3 & 6 \\ 0 & 6 & 1 \end{bmatrix}$

13. The matrix of the form is $\begin{bmatrix} 3 & -1 \\ -1 & 3 \end{bmatrix}$ which is orthogonally diagonalized by

$$Q = \begin{bmatrix} \frac{1}{\sqrt{2}} & -\frac{1}{\sqrt{2}} \\ \frac{1}{\sqrt{2}} & \frac{1}{\sqrt{2}} \end{bmatrix}, \quad D = \begin{bmatrix} 2 & 0 \\ 0 & 4 \end{bmatrix}$$

Let $\mathbf{y} = (x,', y')$. Then

$$\mathbf{y} = Q^T\mathbf{x} = \begin{bmatrix} \frac{x}{\sqrt{2}} + \frac{y}{\sqrt{2}} \\ -\frac{x}{\sqrt{2}} + \frac{y}{\sqrt{2}} \end{bmatrix}$$

and

$$q(\mathbf{x}) = \mathbf{y}^T \begin{bmatrix} 2 & 0 \\ 0 & 4 \end{bmatrix} \mathbf{y} = 2x'^2 + 4y'^2$$

15. The matrix of the form is $\begin{bmatrix} -4 & 2 \\ 2 & -4 \end{bmatrix}$ which is orthogonally diagonalized by

$$Q = \begin{bmatrix} \frac{1}{\sqrt{2}} & -\frac{1}{\sqrt{2}} \\ \frac{1}{\sqrt{2}} & \frac{1}{\sqrt{2}} \end{bmatrix}, \quad D = \begin{bmatrix} -2 & 0 \\ 0 & -6 \end{bmatrix}$$

Let $\mathbf{y} = (x,', y')$. Then

$$\mathbf{y} = Q^T\mathbf{x} = \begin{bmatrix} \frac{x}{\sqrt{2}} + \frac{y}{\sqrt{2}} \\ -\frac{x}{\sqrt{2}} + \frac{y}{\sqrt{2}} \end{bmatrix}$$

and

$$q(\mathbf{x}) = \mathbf{y}^T \begin{bmatrix} -2 & 0 \\ 0 & -6 \end{bmatrix} \mathbf{y} = -2x'^2 - 6y'^2$$

17. The matrix of the form is $\begin{bmatrix} 2 & 0 & -1 \\ 0 & 0 & 0 \\ -1 & 0 & 2 \end{bmatrix}$ which is orthogonally diagonalized by

$$Q = \begin{bmatrix} 0 & \frac{1}{\sqrt{2}} & -\frac{1}{\sqrt{2}} \\ 1 & 0 & 0 \\ 0 & \frac{1}{\sqrt{2}} & \frac{1}{\sqrt{2}} \end{bmatrix}, \qquad D = \begin{bmatrix} 0 & 0 & 0 \\ 0 & 1 & 0 \\ 0 & 0 & 3 \end{bmatrix}$$

Let $\mathbf{y} = (x,', y', z')$. Then

$$\mathbf{y} = Q^T\mathbf{x} = \begin{bmatrix} y \\ \frac{x}{\sqrt{2}} + \frac{z}{\sqrt{2}} \\ -\frac{x}{\sqrt{2}} + \frac{z}{\sqrt{2}} \end{bmatrix}$$

and

$$q(\mathbf{x}) = \mathbf{y}^T \begin{bmatrix} 0 & 0 & 0 \\ 0 & 1 & 0 \\ 0 & 0 & 3 \end{bmatrix} \mathbf{y} = y'^2 + 3z'^2$$

19. The matrix of the form is $\begin{bmatrix} 5 & -4 & 0 \\ -4 & 3 & 4 \\ 0 & 4 & 1 \end{bmatrix}$ which is orthogonally diagonalized by

$$Q = \begin{bmatrix} \frac{1}{3} & -\frac{2}{3} & \frac{2}{3} \\ \frac{2}{3} & \frac{2}{3} & \frac{1}{3} \\ -\frac{2}{3} & \frac{1}{3} & \frac{2}{3} \end{bmatrix}, \qquad D = \begin{bmatrix} -3 & 0 & 0 \\ 0 & 9 & 0 \\ 0 & 0 & 3 \end{bmatrix}$$

Let $\mathbf{y} = (x,', y', z')$. Then

$$\mathbf{y} = Q^T\mathbf{x} = \begin{bmatrix} \frac{1}{3}x + \frac{2}{3}y - \frac{2}{3}z \\ -\frac{2}{3}x + \frac{2}{3}y + \frac{1}{3}z \\ \frac{2}{3}x + \frac{1}{3}y + \frac{2}{3}z \end{bmatrix}$$

and

$$q(\mathbf{x}) = \mathbf{y}^T \begin{bmatrix} -3 & 0 & 0 \\ 0 & 9 & 0 \\ 0 & 0 & 3 \end{bmatrix} \mathbf{y} = -3x'^2 + 9y'^2 + 3z'^2$$

21. The matrix of the form is $\begin{bmatrix} 5 & -4 \\ -4 & 5 \end{bmatrix}$ which is orthogonally diagonalized by

$$Q = \begin{bmatrix} \frac{1}{\sqrt{2}} & -\frac{1}{\sqrt{2}} \\ \frac{1}{\sqrt{2}} & \frac{1}{\sqrt{2}} \end{bmatrix}, \quad D = \begin{bmatrix} 1 & 0 \\ 0 & 9 \end{bmatrix}.$$

Let $\mathbf{y} = (x,', y')$. Then

$$\mathbf{y} = Q^T\mathbf{x} = \begin{bmatrix} \frac{x}{\sqrt{2}} + \frac{y}{\sqrt{2}} \\ -\frac{x}{\sqrt{2}} + \frac{y}{\sqrt{2}} \end{bmatrix}$$

and

$$q(\mathbf{x}) = \mathbf{y}^T \begin{bmatrix} 1 & 0 \\ 0 & 9 \end{bmatrix} \mathbf{y} = x'^2 + 9y'^2 = 1$$

Hence, the conic section is an ellipse.

23. According to the principal axes theorem, a quadratic form q can be written as

$$q(\mathbf{x}) = \lambda_1 y_1^2 + \cdots + \lambda_n y_n^2$$

by a change of variables. The coefficients $\lambda_1 \ldots, \lambda_n$ are the eigenvalues of the matrix A of q.

1. Suppose that q is positive definite. Hence, for any values of the y_is we have that $q(\mathbf{x}) > 0$. By letting the vector $\mathbf{y} = (y_1, \ldots, y_n)$ run through $\mathbf{e}_1, \ldots, \mathbf{e}_n$ we get $\lambda_1, \ldots, \lambda_n$, so these numbers have to be positive. Hence, the eigenvalues of A are positive.

Conversely, if the eigenvalues of A are all positive $\lambda_1 > 0, \ldots, \lambda_n > 0$, then $q(\mathbf{x}) > 0$, because all $y_i^2 \geq 0$ and not all y_i are zero. (We exclude any trivial forms). Therefore, q is positive definite.

2. The proof of this part is identical to the proof of part 1 except that we change

">" to "<" and "positive" to "negative".

3. Let q be an indefinite quadratic form. Then $q(\mathbf{x})$ takes on both positive and negative values. But if all the $\lambda_i s$ were positive, the form would be positive definite (by part 1). Likewise, if all the $\lambda_i s$ were negative, the form would be negative definite (by part 2). So, we conclude that some of the $\lambda_i s$ are positive and some are negative.

Conversely, let some of the $\lambda_i s$ be positive and some be negative. Then by part 1, the form cannot be positive definite. By part 2 it cannot be negative definite either. So, it must be indefinite.

25. If A is orthogonally diagonalized by Q and D, then $Q^T AQ = D$. Hence,

$$Q^T A^2 Q = (Q^T AQ)(Q^T AQ) = DD = D^2$$

So, Q and D^2 orthogonally diagonalize A^2. But then the eigenvalues of A are the diagonal entries of D^2, hence they are nonnegative. Therefore, A^2 is positive semidefinite by a version of Theorem 29 for semidefinite matrices.

27. By the spectral theorem A can be orthogonally diagonalized, say by Q and D. So, $A = QDQ^T$. Because A is positive definite, all its eigenvalues are positive. Thus, all diagonal entries of D are positive. Let B be the diagonal matrix with diagonal entries the square roots of the diagonal entries of D. So,

$$D = B^2 = BB = B^T B$$

Then

$$A = QDQ^T = QB^T BQ^T$$

Let $P = BQ^T$. Then P is invertible, because B and Q^T are, and $A = P^T P$, as asserted.

29. We have

$$\begin{aligned} q(x,y) &= ax^2 + bxy + cy^2 \\ &= a(x^2 + 2x\frac{b}{2a}y + \frac{c}{a}y^2 + \frac{b^2}{4a^2}y^2 - \frac{b^2}{4a^2}y^2) \\ &= a\left(x + \frac{b}{2a}y\right)^2 + \left(c - \frac{b^2}{4a}\right)y^2 \\ &= aX^2 + By^2 \end{aligned}$$

where, $X = x + \dfrac{b}{2a}y$ and $B = c - \dfrac{b^2}{4a}$.

31. Yes, as follows:

$$\begin{aligned} q(x,y) &= bxy + cy^2 \\ &= c(y^2 + 2\frac{b}{2c}xy + \frac{b^2}{4c^2}x^2 - \frac{b^2}{4c^2}x^2) \\ &= c\left(y + \frac{b}{2c}x\right)^2 + \frac{-b^2}{4c}x^2 \\ &= cY^2 + Ax^2 \end{aligned}$$

where, $Y = y + \dfrac{b}{2c}x$ and $A = \dfrac{-b^2}{4c^2}$.

Exercises 8.7

---

1. Because $A^TA = \begin{bmatrix} 9 & 0 \\ 0 & 4 \end{bmatrix}$ the singular values are $3, 2$.

3. Because $A^TA = \begin{bmatrix} 2 & 0 & 2 \\ 0 & 1 & 0 \\ 2 & 0 & 2 \end{bmatrix}$ has eigenvalues $4, 1, 0$, the singular values of A are $2, 1, 0$.

5. $A^TA = \begin{bmatrix} 4 & 0 & 0 \\ 0 & 0 & 0 \\ 0 & 0 & 25 \end{bmatrix}$ has eigenvalues $25, 4, 0$ with corresponding eigenvectors $(0,0,1)$, $(1,0,0)$, and $(0,1,0,)$. Hence, the singular values are $5, 2, 0$ and $V = \begin{bmatrix} 0 & 1 & 0 \\ 0 & 0 & 1 \\ 1 & 0 & 0 \end{bmatrix}$. Therefore,

$$\mathbf{u}_1 = \frac{1}{5}\begin{bmatrix} -2 & 0 & 0 \\ 0 & 0 & 5 \end{bmatrix}\begin{bmatrix} 0 \\ 0 \\ 1 \end{bmatrix} = \begin{bmatrix} 0 \\ 1 \end{bmatrix}, \quad \mathbf{u}_2 = \frac{1}{2}\begin{bmatrix} -2 & 0 & 0 \\ 0 & 0 & 5 \end{bmatrix}\begin{bmatrix} 1 \\ 0 \\ 0 \end{bmatrix} = \begin{bmatrix} -1 \\ 0 \end{bmatrix}$$

Thus, an SVD of A is

$$\begin{bmatrix} 0 & -1 \\ 1 & 0 \end{bmatrix} \begin{bmatrix} 5 & 0 & 0 \\ 0 & 2 & 0 \end{bmatrix} \begin{bmatrix} 0 & 1 & 0 \\ 0 & 0 & 1 \\ 1 & 0 & 0 \end{bmatrix}^T$$

7. Because $A^TA = \begin{bmatrix} 9 & 0 & 0 \\ 0 & 4 & 0 \\ 0 & 0 & 1 \end{bmatrix}$, the singular values are $3, 2, 1$ and $V = I_3$. Therefore,

$$\mathbf{u}_1 = \frac{1}{3}\begin{bmatrix} 0 & 0 & 1 \\ 0 & 2 & 0 \\ 3 & 0 & 0 \end{bmatrix}\begin{bmatrix} 1 \\ 0 \\ 0 \end{bmatrix} = \begin{bmatrix} 0 \\ 0 \\ 1 \end{bmatrix}, \qquad \mathbf{u}_2 = \frac{1}{2}\begin{bmatrix} 0 & 0 & 1 \\ 0 & 2 & 0 \\ 3 & 0 & 0 \end{bmatrix}\begin{bmatrix} 0 \\ 1 \\ 0 \end{bmatrix} = \begin{bmatrix} 0 \\ 1 \\ 0 \end{bmatrix},$$

$$\mathbf{u}_3 = \frac{1}{1}\begin{bmatrix} 0 & 0 & 1 \\ 0 & 2 & 0 \\ 3 & 0 & 0 \end{bmatrix}\begin{bmatrix} 0 \\ 0 \\ 1 \end{bmatrix} = \begin{bmatrix} 1 \\ 0 \\ 0 \end{bmatrix}$$

Thus, an SVD of A is

$$\begin{bmatrix} 0 & 0 & 1 \\ 0 & 1 & 0 \\ 1 & 0 & 0 \end{bmatrix} \begin{bmatrix} 3 & 0 & 0 \\ 0 & 2 & 0 \\ 0 & 0 & 1 \end{bmatrix} \begin{bmatrix} 1 & 0 & 0 \\ 0 & 1 & 0 \\ 0 & 0 & 1 \end{bmatrix}^T$$

9. $A^TA = \begin{bmatrix} 20 & 0 & 40 \\ 0 & 16 & 0 \\ 40 & 0 & 80 \end{bmatrix}$ has eigenvalues $100, 16, 0$ with corresponding eigenvectors $(1, 0, 2), (0, 1, 0)$, and $(-2, 0, 1)$. Hence, the singular values are $10, 4, 0$, and after orthogonalization, $V = \begin{bmatrix} \frac{1}{\sqrt{5}} & 0 & -\frac{2}{\sqrt{5}} \\ 0 & 1 & 0 \\ \frac{2}{\sqrt{5}} & 0 & \frac{1}{\sqrt{5}} \end{bmatrix}$.

$$\mathbf{u}_1 = \frac{1}{10}\begin{bmatrix} 2 & 0 & 4 \\ 0 & 4 & 0 \\ 4 & 0 & 8 \end{bmatrix}\begin{bmatrix} \frac{1}{\sqrt{5}} \\ 0 \\ \frac{2}{\sqrt{5}} \end{bmatrix} = \begin{bmatrix} \frac{1}{\sqrt{5}} \\ 0 \\ \frac{2}{\sqrt{5}} \end{bmatrix}$$

$$\mathbf{u}_2 = \frac{1}{4}\begin{bmatrix} 2 & 0 & 4 \\ 0 & 4 & 0 \\ 4 & 0 & 8 \end{bmatrix}\begin{bmatrix} 0 \\ 1 \\ 0 \end{bmatrix} = \begin{bmatrix} 0 \\ 1 \\ 0 \end{bmatrix}$$

and we extend to get $\mathbf{u}_3 = \begin{bmatrix} -\frac{2}{\sqrt{5}} \\ 0 \\ \frac{1}{\sqrt{5}} \end{bmatrix}$. Thus, an SVD of A is

$$\begin{bmatrix} \frac{1}{\sqrt{5}} & 0 & -\frac{2}{\sqrt{5}} \\ 0 & 1 & 0 \\ \frac{2}{\sqrt{5}} & 0 & \frac{1}{\sqrt{5}} \end{bmatrix} \begin{bmatrix} 10 & 0 & 0 \\ 0 & 4 & 0 \\ 0 & 0 & 0 \end{bmatrix} \begin{bmatrix} \frac{1}{\sqrt{5}} & 0 & -\frac{2}{\sqrt{5}} \\ 0 & 1 & 0 \\ \frac{2}{\sqrt{5}} & 0 & \frac{1}{\sqrt{5}} \end{bmatrix}^T$$

11. Because $A^TA = \begin{bmatrix} 36 & 0 & 0 \\ 0 & 81 & 0 \\ 0 & 0 & 36 \end{bmatrix}$, the singular values are $9, 6, 6$ and $V = \begin{bmatrix} 0 & 1 & 0 \\ 1 & 0 & 0 \\ 0 & 0 & 1 \end{bmatrix}$.

Therefore,

$$\mathbf{u}_1 = \frac{1}{9}\begin{bmatrix} 2 & 6 & -4 \\ -4 & 6 & 2 \\ 4 & 3 & 4 \end{bmatrix}\begin{bmatrix} 0 \\ 1 \\ 0 \end{bmatrix} = \begin{bmatrix} \frac{2}{3} \\ \frac{2}{3} \\ \frac{1}{3} \end{bmatrix}$$

$$\mathbf{u}_2 = \frac{1}{6}\begin{bmatrix} 2 & 6 & -4 \\ -4 & 6 & 2 \\ 4 & 3 & 4 \end{bmatrix}\begin{bmatrix} 1 \\ 0 \\ 0 \end{bmatrix} = \begin{bmatrix} \frac{1}{3} \\ -\frac{2}{3} \\ \frac{2}{3} \end{bmatrix}$$

$$\mathbf{u}_3 = \frac{1}{6}\begin{bmatrix} 2 & 6 & -4 \\ -4 & 6 & 2 \\ 4 & 3 & 4 \end{bmatrix}\begin{bmatrix} 0 \\ 0 \\ 1 \end{bmatrix} = \begin{bmatrix} -\frac{2}{3} \\ \frac{1}{3} \\ \frac{2}{3} \end{bmatrix}$$

Thus, an SVD of A is

$$\begin{bmatrix} \frac{2}{3} & \frac{1}{3} & -\frac{2}{3} \\ \frac{2}{3} & -\frac{2}{3} & \frac{1}{3} \\ \frac{1}{3} & \frac{2}{3} & \frac{2}{3} \end{bmatrix} \begin{bmatrix} 9 & 0 & 0 \\ 0 & 6 & 0 \\ 0 & 0 & 6 \end{bmatrix} \begin{bmatrix} 0 & 1 & 0 \\ 1 & 0 & 0 \\ 0 & 0 & 1 \end{bmatrix}^T$$

13. A^TA is messy, but AA^T is diagonal. An SVD of A^T is

$$\begin{bmatrix} \frac{2}{3} & \frac{2}{3} & \frac{1}{3} \\ -\frac{2}{3} & \frac{1}{3} & \frac{2}{3} \\ \frac{1}{3} & -\frac{2}{3} & \frac{2}{3} \end{bmatrix} \begin{bmatrix} 9 & 0 & 0 \\ 0 & 3 & 0 \\ 0 & 0 & 0 \end{bmatrix} \begin{bmatrix} 0 & 1 & 0 \\ 0 & 0 & 1 \\ 1 & 0 & 0 \end{bmatrix}^T$$

Hence, an SVD of A is obtained by transposition:

$$\begin{bmatrix} 0 & 1 & 0 \\ 0 & 0 & 1 \\ 1 & 0 & 0 \end{bmatrix} \begin{bmatrix} 9 & 0 & 0 \\ 0 & 3 & 0 \\ 0 & 0 & 0 \end{bmatrix} \begin{bmatrix} \frac{2}{3} & \frac{2}{3} & \frac{1}{3} \\ -\frac{2}{3} & \frac{1}{3} & \frac{2}{3} \\ \frac{1}{3} & -\frac{2}{3} & \frac{2}{3} \end{bmatrix}^T$$

15. (a) The verification of the Moore-Penrose properties is straightforward given that

$$A^+ = \begin{bmatrix} 0 & 0 & 1 \\ 0 & 1 & 0 \\ 1 & 0 & 0 \end{bmatrix} \begin{bmatrix} \frac{1}{6} & 0 & 0 \\ 0 & \frac{1}{4} & 0 \\ 0 & 0 & \frac{1}{2} \end{bmatrix} \begin{bmatrix} 0 & 0 & 1 \\ 0 & 1 & 0 \\ 1 & 0 & 0 \end{bmatrix}^T = \begin{bmatrix} \frac{1}{2} & 0 & 0 \\ 0 & \frac{1}{4} & 0 \\ 0 & 0 & \frac{1}{6} \end{bmatrix}$$

(b) The verification of the Moore-Penrose properties is straightforward given that

$$A^+ = \begin{bmatrix} 1 & 0 & 0 \\ 0 & 1 & 0 \\ 0 & 0 & 1 \end{bmatrix} \begin{bmatrix} \frac{1}{3} & 0 & 0 \\ 0 & \frac{1}{2} & 0 \\ 0 & 0 & 1 \end{bmatrix} \begin{bmatrix} 0 & 0 & 1 \\ 0 & 1 & 0 \\ 1 & 0 & 0 \end{bmatrix}^T = \begin{bmatrix} 0 & 0 & \frac{1}{3} \\ 0 & \frac{1}{2} & 0 \\ 1 & 0 & 0 \end{bmatrix}$$

17. A^+ was computed in Exercise 15(b) and was found to be $\begin{bmatrix} 0 & 0 & \frac{1}{3} \\ 0 & \frac{1}{2} & 0 \\ 1 & 0 & 0 \end{bmatrix}$ which is also A^{-1}.

19. We have

$$AA^+A = \begin{bmatrix} -3 & 0 \\ 0 & 0 \\ 0 & 4 \end{bmatrix}, \quad A^+AA^+ = \begin{bmatrix} -\frac{1}{3} & 0 & 0 \\ 0 & 0 & \frac{1}{4} \end{bmatrix}$$

$$(AA^+)^T = \begin{bmatrix} 1 & 0 & 0 \\ 0 & 0 & 0 \\ 0 & 0 & 1 \end{bmatrix}, \quad AA^+ = \begin{bmatrix} 1 & 0 & 0 \\ 0 & 0 & 0 \\ 0 & 0 & 1 \end{bmatrix}$$

$$(A^+A)^T = I, \qquad A^+A = I$$

21. It suffices to verify the Moore-Penrose conditions for (A^+, A). Because then A would be the unique inverse of A^+. So, $A^{++} = A$. By Exercise 18, we have

$$A^+AA^+ = A^+, \;\; AA^+A = A, \;\; (A^+A)^T = A^TA, \;\; (AA^+)^T = AA^+$$

Hence, the conditions hold for the pair (A^+, A).

23. $A^{+}\mathbf{b} = \begin{bmatrix} -\frac{1}{2} & 0 & 0 \\ 0 & 0 & \frac{1}{5} \end{bmatrix} \begin{bmatrix} 1 \\ 2 \\ 3 \end{bmatrix} = \begin{bmatrix} -\frac{1}{2} \\ \frac{3}{5} \end{bmatrix}$

25. $A^{+}\mathbf{b} = \begin{bmatrix} -\frac{2}{9} & \frac{1}{9} & \frac{2}{9} \\ \frac{2}{27} & \frac{2}{27} & \frac{1}{27} \end{bmatrix} \begin{bmatrix} 1 \\ 2 \\ 3 \end{bmatrix} = \begin{bmatrix} \frac{2}{3} \\ \frac{1}{3} \end{bmatrix}$

27. (a) $A = PQ = \begin{bmatrix} \sqrt{2} & 0 \\ 0 & \sqrt{2} \end{bmatrix} \begin{bmatrix} \frac{1}{\sqrt{2}} & -\frac{1}{\sqrt{2}} \\ \frac{1}{\sqrt{2}} & \frac{1}{\sqrt{2}} \end{bmatrix}$

(b) $A = PQ = \begin{bmatrix} 7 & 2 & 0 \\ 2 & 6 & 2 \\ 0 & 2 & 5 \end{bmatrix} \begin{bmatrix} \frac{1}{3} & \frac{2}{3} & -\frac{2}{3} \\ -\frac{2}{3} & \frac{2}{3} & \frac{1}{3} \\ \frac{2}{3} & \frac{1}{3} & \frac{2}{3} \end{bmatrix}$

Exercises 8.8

---

1. (a) $-12, \quad \sqrt{19}, \quad 2\sqrt{21}, \quad \sqrt{79}$ (b) $\|(-3,-5)\| = \sqrt{127}$

(c) $\sqrt{19}\cdot 2\sqrt{21} \simeq 39.95 \geq |-12| = 12$ (d) $\sqrt{79} \simeq 8.89 \leq \sqrt{19} + 2\sqrt{21} \simeq 13.52$

(e) $\frac{1}{4}(79 - 127) = -12$

3. We need numbers x and y such that

$$\begin{bmatrix} 1 \\ -2 \end{bmatrix}^{T} \begin{bmatrix} 3 & 0 \\ 0 & 4 \end{bmatrix} \begin{bmatrix} x \\ y \end{bmatrix} = 3x - 8y = 0$$

For example, $\left(1, \frac{3}{8}\right)$ is orthogonal to $\mathbf{u}$.

5. A and B are orthogonal. Also A and C are orthogonal.

7. The orthonormal pairs are (A', B') and (A', C'), where

$$A' = \frac{1}{\|A\|}A = \begin{bmatrix} \frac{1}{\sqrt{2}} & 0 \\ 0 & \frac{1}{\sqrt{2}} \end{bmatrix}, \qquad B' = \frac{1}{\|B\|}B = \begin{bmatrix} -\frac{1}{\sqrt{2}} & 0 \\ 0 & \frac{1}{\sqrt{2}} \end{bmatrix},$$

$$C' = \frac{1}{\|C\|}C = C = \begin{bmatrix} -\frac{1}{2} & \frac{1}{2} \\ \frac{1}{2} & \frac{1}{2} \end{bmatrix}$$

9. f defines an inner product, because A is positive definite.

11. f defines an inner product, because A is positive definite.

13. f does not define an inner product, because it is not positive definite. For example, $f(\mathbf{u}, \mathbf{u}) = 0$ for $\mathbf{u} = (-1, 1)$.

15. f does not define an inner product, because it is not symmetric. For example, for $\mathbf{u} = (1, 2)$ and $\mathbf{v} = (2, 1)$ we have $f(\mathbf{u}, \mathbf{v}) = 31$, whereas $f(\mathbf{v}, \mathbf{u}) = 49$.

17. It suffices to check the four axioms of the definition of the inner product.

1. The transpose of a single number as a 1×1 matrix is the number itself, so we have:

$$\langle \mathbf{u}, \mathbf{v} \rangle = \mathbf{v}^T A\mathbf{u} = \left(\mathbf{v}^T A\mathbf{u}\right)^T = \mathbf{u}^T A^T \mathbf{v}^{TT} = \mathbf{u}^T A\mathbf{v} = \langle \mathbf{v}, \mathbf{u} \rangle$$

because $A^T = A$.

2. Also,

$$\langle \mathbf{u} + \mathbf{w}, \mathbf{v} \rangle = \mathbf{v}^T A(\mathbf{u} + \mathbf{w}) = \mathbf{v}^T(A\mathbf{u} + A\mathbf{w}) = \mathbf{v}^T A\mathbf{u} + \mathbf{v}^T A\mathbf{w} = \langle \mathbf{u}, \mathbf{v} \rangle + \langle \mathbf{w}, \mathbf{v} \rangle$$

3. We have

$$\langle c\mathbf{u}, \mathbf{v} \rangle = \mathbf{v}^T A(c\mathbf{u}) = c\mathbf{v}^T A\mathbf{u} = c \langle \mathbf{u}, \mathbf{v} \rangle$$

4. Also,

$$\langle \mathbf{u}, \mathbf{u} \rangle = \mathbf{u}^T A\mathbf{u} > 0$$

for nonzero $\mathbf{u}$, because A is positive definite.

19. (a) $-255, \ \sqrt{339}, \ 15, \ 3\sqrt{6}$ (b) $\|1 - 3x^2 + 2x\| = \sqrt{1074}$

(c) $|-255| = 255 \leq \sqrt{339} \cdot 15 \simeq 276.18$ (d) $3\sqrt{6} \simeq 7.3485 \leq \sqrt{339} + 15 \simeq 33.412$

21. $2 + x^2$

23. Let $w(x) = a + bx + cx^2$ such that

$$\int_{-1}^{1} p(x)\left(a + bx + cx^2\right) dx = -\frac{2}{15}c + \frac{2}{3}a = 0$$

Thus, we may take $w(x) = \frac{1}{5} + x^2$.

25. 2. We have

$$\langle \mathbf{u}, c\mathbf{v}\rangle = \langle c\mathbf{v}, \mathbf{u}\rangle = c\langle \mathbf{v}, \mathbf{u}\rangle = c\langle \mathbf{u}, \mathbf{v}\rangle$$

3. Also,

$$\begin{aligned}\langle \mathbf{u} - \mathbf{w}, \mathbf{v}\rangle &= \langle \mathbf{u} + (-\mathbf{w}), \mathbf{v}\rangle \\ &= \langle \mathbf{u}, \mathbf{v}\rangle + \langle (-1)\mathbf{w}, \mathbf{v}\rangle \\ &= \langle \mathbf{u}, \mathbf{v}\rangle + (-1)\langle \mathbf{w}, \mathbf{v}\rangle \\ &= \langle \mathbf{u}, \mathbf{v}\rangle - \langle \mathbf{w}, \mathbf{v}\rangle\end{aligned}$$

4. By part 3, we have

$$\langle \mathbf{u}, \mathbf{v} - \mathbf{w}\rangle = \langle \mathbf{v} - \mathbf{w}, \mathbf{u}\rangle = \langle \mathbf{v}, \mathbf{u}\rangle - \langle \mathbf{w}, \mathbf{u}\rangle = \langle \mathbf{u}, \mathbf{v}\rangle + \langle \mathbf{u}, \mathbf{w}\rangle$$

5. We have

$$\langle \mathbf{0}, \mathbf{v}\rangle = \langle 0\mathbf{0}, \mathbf{v}\rangle = 0\langle \mathbf{0}, \mathbf{v}\rangle = 0$$

27. We need to check the axioms.

1. We have

$$\langle \mathbf{u}, \mathbf{v}\rangle = T(\mathbf{u}) \cdot T(\mathbf{v}) = T(\mathbf{v}) \cdot T(\mathbf{u}) = \langle \mathbf{v}, \mathbf{u}\rangle$$

2. Also,

$$\begin{aligned}\langle \mathbf{u} + \mathbf{w}, \mathbf{v}\rangle &= T(\mathbf{u} + \mathbf{w}) \cdot T(\mathbf{v}) \\ &= (T(\mathbf{u}) + T(\mathbf{w})) \cdot T(\mathbf{v}) \\ &= T(\mathbf{u}) \cdot T(\mathbf{v}) + T(\mathbf{w}) \cdot T(\mathbf{v}) \\ &= \langle \mathbf{u}, \mathbf{v}\rangle + \langle \mathbf{w}, \mathbf{v}\rangle\end{aligned}$$

3. We also have

$$\langle c\mathbf{u}, \mathbf{v}\rangle = T(c\mathbf{u}) \cdot T(\mathbf{v}) = cT(\mathbf{v}) \cdot T(\mathbf{u}) = c\langle \mathbf{v}, \mathbf{u}\rangle$$

4. If $\mathbf{u}$ is nonzero so is $T(\mathbf{u})$ because T is invertible. Hence,

$$\langle \mathbf{u}, \mathbf{u}\rangle = T(\mathbf{u}) \cdot T(\mathbf{u}) > 0$$

29. Using the calculation $(1,1)_{\mathrm{pr}} = \left(\frac{2}{3},0\right)$ from the previous exercise the distance from $(1,1)$ to l is

$$\|(1,1)_{\mathrm{c}}\| = \|(1,1)-(1,1)_{\mathrm{pr}}\| = \left\|(1,1)-\left(\frac{2}{3},0\right)\right\| = \left\|\left(\frac{1}{3},1\right)\right\| = \sqrt{\frac{7}{3}}$$

31. Let $f(x) = x^2+x+1$. From the previous exercise we have $f_{\mathrm{pr}} = x^2+x-\frac{8}{3}$. Hence, the distance from x^2+x+1 to W is

$$\|f_{\mathrm{c}}\| = \|f - f_{\mathrm{pr}}\| = \left\|\frac{11}{3}\right\| \doteq \frac{11}{\sqrt{3}}$$

33. Let $\mathbf{u}_1 = 1$. Then

$$\mathbf{u}_2 = x - \frac{0}{3}1 = x, \qquad \mathbf{u}_3 = x^2 - \frac{8}{3}1 - \frac{0}{4}x = x^2 - \frac{8}{3}$$

So, $\left\{1, x, x^2-\frac{8}{3}\right\}$ is an orthogonal basis of P_2 for this inner product.

35. Because

$$\int_{-1}^{1} 1^2 dx = 2$$
$$\int_{-1}^{1} x^2 dx = \frac{2}{3}$$
$$\int_{-1}^{1} \left(\frac{3}{2}x^2 - \frac{1}{2}\right)^2 dx = \frac{2}{5}$$
$$\int_{-1}^{1} \left(\frac{5}{2}x^3 - \frac{3}{2}x\right)^2 dx = \frac{2}{7}$$

an orthonormal basis for P_3 is

$$\mathcal{L}' = \left\{\frac{1}{2}, \frac{3}{2}x, \frac{15}{4}x^2 - \frac{5}{4}, \frac{35}{4}x^3 - \frac{21}{4}x\right\}$$

Exercises 8.9

1. This time we get the solution

$$\mathbf{x} = (1.9847,\ \ 0.8318,\ \ 3.3135,\ \ -4.1813,\ \ 4.2094)$$

which differs from the one discussed in the text by

$$(0.0745,\ \ 0.0003,\ \ 0.0044,\ \ 0.0147,\ \ -0.0211)$$

The answer is satisfactory but we do not quite get the correct rational coefficients.

3. The computed normal equations

$$\begin{bmatrix} 5 & 15 & 55 \\ 15 & 55 & 225 \\ 55 & 225 & 979 \end{bmatrix} \tilde{\mathbf{x}} = \begin{bmatrix} 34 \\ 161 \\ 741 \end{bmatrix}$$

yield $\tilde{\mathbf{x}} = \left(-\frac{22}{5}, \frac{23}{70}, \frac{13}{14}\right)$, thus $q(x) = -\dfrac{22}{5} + \dfrac{23}{70}x + \dfrac{13}{14}x^2$ and $q(6) = 31$.

5. The computed normal equations

$$\begin{bmatrix} 5 & 0 & 10 \\ 0 & 10 & 0 \\ 10 & 0 & 34 \end{bmatrix} \tilde{\mathbf{x}} = \begin{bmatrix} 22 \\ 20 \\ 48 \end{bmatrix}$$

yield $\tilde{\mathbf{x}} = \left(\frac{134}{35}, 2, \frac{2}{7}\right)$. Thus, $q(x) = \dfrac{134}{35} + 2x + \dfrac{2}{7}x^2$ and $q(6) = \dfrac{914}{35} = 26.114$.

7. The computed normal equations

$$\begin{bmatrix} 5 & 0 & 10 & 0 \\ 0 & 10 & 0 & 34 \\ 10 & 0 & 34 & 0 \\ 0 & 34 & 0 & 130 \end{bmatrix} \tilde{\mathbf{x}} = \begin{bmatrix} 3 \\ 15 \\ 17 \\ 69 \end{bmatrix}$$

yield $\tilde{\mathbf{x}} = \left(-\frac{34}{35}, -\frac{11}{4}, \frac{11}{14}, \frac{5}{4}\right)$. Thus, $q(x) = -\dfrac{34}{35} - \dfrac{11}{4}x + \dfrac{11}{14}x^2 + \dfrac{5}{4}x^3$. Also, $q(3) = \dfrac{158}{5} \simeq 31.6$.

9. In the case of n points and an $n-1$ degree least squares polynomial, the matrix A is square of the form

$$A = \begin{bmatrix} 1 & a_1 & a_1^2 & \cdots & a_1^{n-1} \\ \vdots & \vdots & \vdots & & \vdots \\ 1 & a_n & a_n^2 & \cdots & a_n^{n-1} \end{bmatrix}$$

Now A^T is a Vandermonde matrix and its determinant from Section 6.6 is

$$\det(A^T) = \prod_{1 \le i < j \le n} (a_i - a_j)$$

This is a nonzero number because the a_i's are distinct. Hence, $\det(A) \neq 0$ and A is invertible. Thus, the system $\mathbf{b} = A\alpha$ has a unique solution $(\alpha_0, \ldots, \alpha_{n-1})$. So, $q(x) = \alpha_0 + \cdots + \alpha_{n-1}x^{n-1}$ is the unique polynomial such that $q(a_i) = b_i$ for $i = 1, \ldots, n$ and the error vector Δ in the least squares problem is zero.

11. Because

$$\int_1^2 1dx = 1, \quad \int_1^2 xdx = \frac{3}{2}, \quad \int_1^2 x^2dx = \frac{7}{3}, \quad \int_1^2 x^3dx = \frac{15}{4}$$

we have

$$\begin{bmatrix} 1 & \frac{3}{2} & \frac{7}{3} \\ \frac{3}{2} & \frac{7}{3} & \frac{15}{4} \end{bmatrix} \sim \begin{bmatrix} 1 & 0 & -\frac{13}{6} \\ 0 & 1 & 3 \end{bmatrix}$$

So, the line is $y = -\dfrac{13}{6} + 3x$.

13. Because

$$\int_0^2 1dx = 2, \quad \int_0^2 xdx = 2, \quad \int_0^2 x^2dx = \frac{8}{3}, \quad \int_0^2 x^3dx = 4, \quad \int_0^2 x^4dx = \frac{32}{5}$$

we have

$$\begin{bmatrix} 2 & 2 & 4 \\ 2 & \frac{8}{3} & \frac{32}{5} \end{bmatrix} \sim \begin{bmatrix} 1 & 0 & -\frac{8}{5} \\ 0 & 1 & \frac{18}{5} \end{bmatrix}$$

So, the line is $y = -\dfrac{8}{5} + \dfrac{18}{5}x$.

15. Because

$$\int_0^2 1dx = 2, \int_0^2 xdx = 2, \int_0^2 x^2dx = \frac{8}{3}$$

$$\int_0^2 x^3dx = 4, \int_0^2 x^4dx = \frac{32}{5}, \int_0^2 x^5dx = \frac{32}{3}$$

we have

$$\begin{bmatrix} 2 & 2 & \frac{8}{3} & 4 \\ 2 & \frac{8}{3} & 4 & \frac{32}{5} \\ \frac{8}{3} & 4 & \frac{32}{5} & \frac{32}{3} \end{bmatrix} \sim \begin{bmatrix} 1 & 0 & 0 & \frac{2}{5} \\ 0 & 1 & 0 & -\frac{12}{5} \\ 0 & 0 & 1 & 3 \end{bmatrix}$$

So, the quadratic is $y = \frac{2}{5} - \frac{12}{5}x + 3x^2$.

17. We have

$$\langle 1, \sin nx \rangle = \int_{-\pi}^{\pi} 1 \cdot \sin nx dx = -\frac{\cos nx}{n}\Big|_{-\pi}^{\pi} = 0$$

19. We have

$$\begin{aligned} \langle \sin mx, \sin nx \rangle &= \int_{-\pi}^{\pi} \sin mx \sin nx dx \\ &= \frac{1}{2}\frac{\sin(m-n)x}{m-n} - \frac{1}{2}\frac{\sin(m+n)x}{m+n}\Big|_{-\pi}^{\pi} = 0 \end{aligned}$$

21. We have

$$\begin{aligned} a_0 &= \frac{1}{2\pi}\int_0^{\pi} 1dx = \frac{1}{2} \\ a_n &= \frac{1}{\pi}\int_0^{\pi} 1\cos(nx)dx = \frac{1}{\pi}\frac{\sin n\pi}{n} = 0 \\ b_n &= \frac{1}{\pi}\int_0^{\pi} 1\sin(nx)dx = -\frac{1}{\pi}\frac{\cos n\pi - 1}{n} = \frac{1-(-1)^n}{n\pi} \end{aligned}$$

23. We have

$$\begin{aligned} \langle \sin nx, \sin mx \rangle &= \int_0^{\pi} \sin nx \sin mx\, dx \\ &= \frac{1}{2}\frac{\sin(n-m)x}{n-m} - \frac{1}{2}\frac{\sin(n+m)x}{n+m}\Big|_0^{\pi} = 0 \end{aligned}$$

25. We have

$$\begin{aligned}\langle \sin n\pi x, \sin m\pi x\rangle &= \int_{-1}^{1} \sin n\pi x \sin m\pi x \, dx \\ &= \frac{1}{2}\frac{\sin(n\pi - m\pi)x}{n\pi - m\pi} - \frac{1}{2}\frac{\sin(n\pi + m\pi)x}{n\pi + m\pi}\bigg|_{-1}^{1} = 0\end{aligned}$$

27. We have

$$\langle 1, \cos m\pi x\rangle = \int_0^2 \cos m\pi x \, dx = \frac{\sin m\pi x}{m\pi}\bigg|_0^2 = 0$$

and

$$\begin{aligned}\langle \cos n\pi x, \cos m\pi x\rangle &= \int_0^2 \cos n\pi x \cos m\pi x \, dx \\ &= \frac{1}{2}\frac{\sin(n\pi - m\pi)x}{n\pi - m\pi} + \frac{1}{2}\frac{\sin(n\pi + m\pi)x}{n\pi + m\pi}\bigg|_0^2 = 0\end{aligned}$$

29. Because for $m = 1, 2, \dots$

$$\psi_{m,0} = \begin{cases} 2^{-m/2} & 0 \le x \le 2^{m-1} \\ -2^{-m/2} & 2^{m-1} < x \le 2^m \end{cases}$$

we have

$$c_{m,0} = \int_{-\infty}^{\infty} f(x)\psi_{m,0}(x)dx = \int_0^1 (-1)\psi_{m,0}(x)dx = \int_0^1 (-2^{-m/2})dx = -2^{-m/2}$$

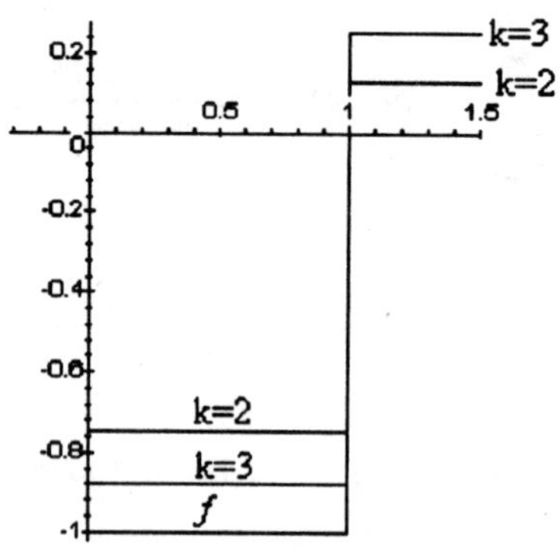